A-Z NANOBIOTECHNOLOGY

A-Z NANOBIOTECHNOLOGY

Dr. Albert Shawn

CENTRUM PRESS
NEW DELHI-110002 (INDIA)

CENTRUM PRESS
H.O.: 4360/4, Ansari Road, Daryaganj,
New Delhi-110 002 (India)
Ph.: 23278000, 23261597
B.O.: No. 1015, Ist Main Road, BSK IIIrd Stage
IIIrd Phase, IIIrd Block,
Bangalore - 560 085 (India)
Tel.: 080-41723429
Visit us at: www.centrumpress.com

A-Z Nanobiotechnology

First Edition, 2009
ISBN 978-93-80106-46-5

PRINTED IN INDIA

Printed at Salasar Imaging Systems, Delhi-110035 (India)

Contents

Preface

Nanobiotechnology is the branch of nanotechnology with biological and biochemical applications or uses. Nanobiotechnology often studies existing elements of nature in order to fabricate new devices. The term bionanotechnology is often used interchangeably with nanobiotechnology, though a distinction is sometimes drawn between the two. If the two are distinguished, nanobiotechnology usually refers to the use of nanotechnology to further the goals of biotechnology, while bionanotechnology might refer to any overlap between biology and nanotechnology, including the use of biomolecules as part of or as an inspiration for nanotechnological devices.

The application of nanotechnology in biomedical sciences will provide information with unprecedented precision and sensitivity, which will not only provide much deeper understanding of biosystems but also lead to the development of new revolutionary modalities of biomolecular manufacturing, early diagnostics, medical treatment, and disease prevention beyond the cellular level to that of DNA and individual proteins, the building blocks of the life process. The long-term goal of Nanobiotechnology is to contribute to the information transfer and explosive growth of this important research area. Nanobiotechnology is a rapidly advancing area of scientific and technological opportunity that applies the tools and processes of nano/microfabrication to build devices for studying biosystems. Researchers learn from biology to create new micro-nanoscale devices to better understand life processes at the nanoscale.

Author

PROTEIN-DNA INTERACTIONS

The [illegible] method has shown that the distribution of Cα position of amino acids around base pairs provides important information about the specificity in the DNA sequence recognition by proteins. However, the accuracy of this prediction method is limited by the number of available structural data. Thus it is desirable to complement the method by some other means. Computer simulation [illegible] method [illegible]

[illegible]

[illegible] in the free energy contour maps for the interactions of Asn with A and with G [illegible] the preferable position of Asn is localized in a narrow region around A in the case of A-T. In this region, Asn and A form specific double [illegible]

Chapter 1

Genetic Molecule

The discovery that DNA is the prime genetic molecule, carrying all the hereditary information within chromosomes, immediately focused attention on its structure. It was hoped that knowledge of the structure would reveal how DNA carries the genetic messages that are replicated when chromosomes divide to produce two identical copies of themselves.

During the late 1940s and early 1950s, several research groups in the United States and in Europe engaged in serious efforts—both cooperative and rival—to understand how the atoms of DNA are linked together by covalent bonds and how the resulting molecules are arranged in three-dimensional space. Not surprisingly, there initially were fears that DNA might have very complicated and perhaps bizarre structures that differed radically from one gene to another. Great relief, if not general elation, was thus expressed when the fundamental DNA structure was found to be the double helix.

It told us that all genes have roughly the same three-dimensional form and that the differences between two genes reside in the order and number of their four nucleotide building blocks along the complementary strands. Now, some 50 years after the discovery of the double helix, this simple description of the genetic material remains true and has not had to be appreciably altered to accommodate new findings.

Nevertheless, we have come to realise that the structure of DNA is not quite as uniform as was first thought. For example, the chromosome of some small viruses have single-stranded, not double-stranded, molecules. Moreover, the precise orientation of the base pairs varies slightly from base pair to base pair in a manner that is influenced by the local DNA sequence. Some DNA sequences even permit the double helix to twist in the left-handed sense, as opposed to the right-handed sense originally formulated for DNA's general structure. And while some DNA molecules are linear, others are circular.

Still additional complexity comes from the supercooling (further twisting) of the double helix, often around cores of DNA-binding proteins. Likewise, we now realise that RNA, which at first glance appears to be very similar to DNA, has its own distinctive structural features. It is principally found as a single-stranded molecule.

Yet by means of intra-strand base pairing, RNA exhibits extensive double-helical character and is capable of folding into a wealth of diverse tertiary structures. These structures are full of surprises, such as non-classical base pairs, base-backbone interactions, and knot-like configurations. Most remarkable of all, and of profound evolutionary significance, some RNA molecules are enzymes that carry out reactions that are at the core of information transfer from nucleic acid to protein.

Clearly, the structures of DNA and RNA are richer and more intricate than was at first appreciated. Indeed, there is no one generic structure for DNA and RNA.

DNA Is Composed of Polynucleotide Chains

The most important feature of DNA is that it is usually composed of two polynucleotide chains twisted around each other in the form of a double helix. The upper part of the figure presents the structure of the double helix shown in a schematic form. Note that if inverted 180°, the double helix looks superficially the same, due to the complementary nature of the two DNA strands.

The space-filling model of the double helix, in the lower part of the figure, shows the components of the DNA molecule and their relative positions in the helical structure.

The backbone of each strand of the helix is composed of alternating sugar and phosphate residues; the bases project inward but are accessible through the major and minor grooves.

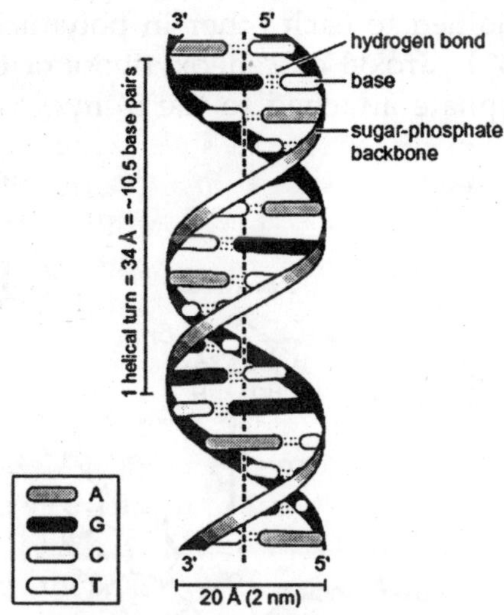

Fig. The Helical Structure of DNA

Let us begin by considering the nature of the nucleotide, the fundamental building block of DNA. The nucleotide consists of a phosphate joined to a sugar, known as 2′-deoxyribose, to which a base is attached. The sugar is called 2′-deoxyribose because there is no hydroxyl at position 2′ (just two hydrogens).

Note that the positions on the ribose are designated with primes to distinguish them from positions on the bases. We can think of how the base is joined to 2′-deoxyribose by imagining the removal of a molecule of water between the hydroxyl on the 1′ carbon of the sugar and the base to form a glycosidic bond.

The sugar and base alone are called a nucleoside. Likewise, we can imagine linking the phosphate to 2′-deoxyribose by removing a water molecule from between the phosphate and the hydroxyl on the 5′ carbon to make a 5′ phosphomonoester. Adding a phosphate (or more than one phosphate) to a nucleo*s*ide creates a nucleo*t*ide. Thus, by making a glycosidic bond between the base and the sugar, and by making a phosphoester bond between the sugar and the phosphoric acid, we have created a nucleotide. Nucleotides are, in turn, joined to each other in polynucleotide chains through the 3′ hydroxyl of 2′-deoxyribose of one nucleotide and the phosphate attached to the 5′ hydroxyl of another nucleotide.

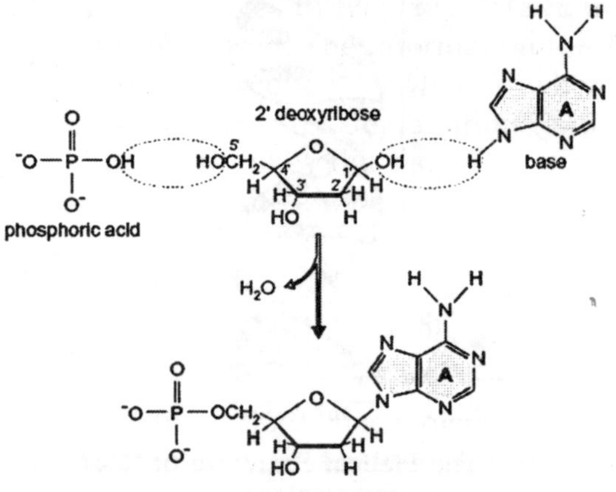

Fig. Formation of Nucleotide by Removal of Water

This is a phosphodiester linkage in which the phosphoryl group between the two nucleotides has one sugar esterified to it through a 3′ hydroxyl and a second sugar esterified to it through a 5′ hydroxyl. Phosphodiester linkages create the repeating, sugar-phosphate backbone of the polynucleotide chain, which is a regular feature of DNA. In contrast, the order of the bases along the polynucleotide chain is irregular. This irregularity as well as the long length is, as we shall see, the basis for the enormous information content of DNA. The phosphodiester linkages impart an inherent polarity to the DNA chain. This polarity is defined by the asymmetry of the nucleotides and the way they are joined. DNA chains have a free 5′ phosphate or 5′ hydroxyl at one end and a free 3′ phosphate or 3′ hydroxyl at the other end. The convention is to write DNA sequences from the 5′ end (on the left) to the 3′ end, generally with a 5′ phosphate and a 3′ hydroxyl.

	Base Adenine	Nucleoside Adenosine	Nucleotide Adenosine 5'-phosphate	Deoxynucleotide Deoxyadenosine 5' phosphate
*Structure**				
M.W.	135.1	267.2	347.2	331.2

EACH BASE HAS ITS PREFERRED TAUTOMERIC FORM

The bases in DNA fall into two classes, purines and pyrimidines. The purines are adenine and guanine, and the pyrimidines are cytosine and thymine. The purines are derived from the double-ringed structure. Adenine and guanine share this essential structure but with different groups attached. Likewise, cytosine and thymine are variations on the single-ringed structure shown in Figure. The numbering of the positions in the purine and pyrimidine rings. The bases are attached to the deoxyribose by glycosidic linkages at N_1 of the pyrimidines or at N9 of the purines.

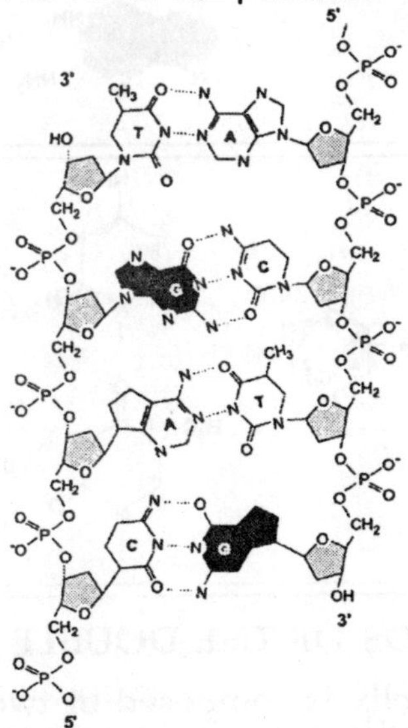

Fig. Detailed Structure of Polynucleotide Polymer

Each of the bases exists in two alternative tautomeric states, which are in equilibrium with each other. The equilibrium lies far to the side of the conventional structures shown in Figure, which are the predominant states and the ones important for base pairing.

The nitrogen atoms attached to the purine and pyrimidine rings are in the amino form in the predominant state and only rarely assume the imino configuration. Likewise, the oxygen atoms attached to the guanine and thymine normally have the keto form and only rarely take on the enol configuration. As examples, tautomerization of cytosine into the imino form (a) and guanine into the enol form (b). As we shall see, the capacity to form an alternative tautomer is a frequent source of errors during DNA synthesis.

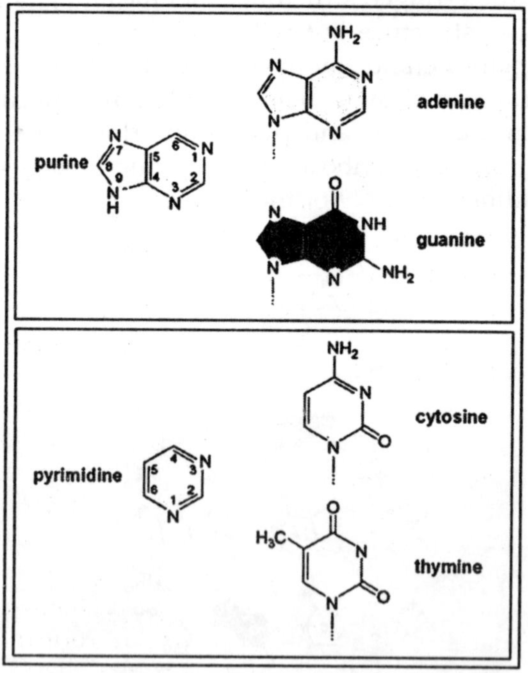

STRANDS OF THE DOUBLE HELIX

The double helix is composed of two polynucleotide chains that are held together by weak, non-covalent bonds between pairs of bases. Adenine on one chain is always paired with thymine on the other chain and, likewise, guanine is always paired with cytosine. The two strands have the same helical geometry but base pairing holds them together with the opposite polarity.

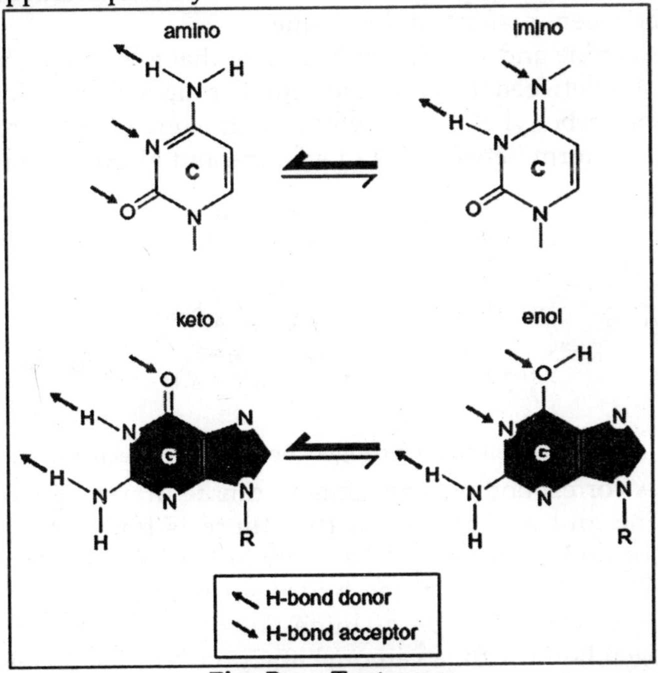

Fig. Base Tautomers

That is, the base at the 5′ end of one strand is paired with the base at the 3′ end of the other strand. The strands are said to have an anti-parallel orientation. This anti-parallel orientation is a stereochemical consequence of the way that adenine and thymine and guanine and cytosine pair with each together.

CHAINS OF THE DOUBLE HELIX

The pairing between adenine and thymine and between guanine and cytosine results in a complementary relationship between the sequence of bases on the two intertwined chains and gives DNA its self-encoding character. For example, if we have the sequence 5′-ATGTC-3′ on one chain, the opposite chain must have the complementary sequence 3′-TACAG-5′. The strictness of the rules for this "Watson-Crick" pairing derives from the complementarity both of shape and of hydrogen bonding properties between adenine and thymine and between guanine and cytosine.

Adenine and thymine match up so that a hydrogen bond can form between the exocyclic amino group at C_6 on adenine and the carbonyl at C_4 in thymine; and likewise, a hydrogen bond can form between N1 of adenine and N_3 of thymine.

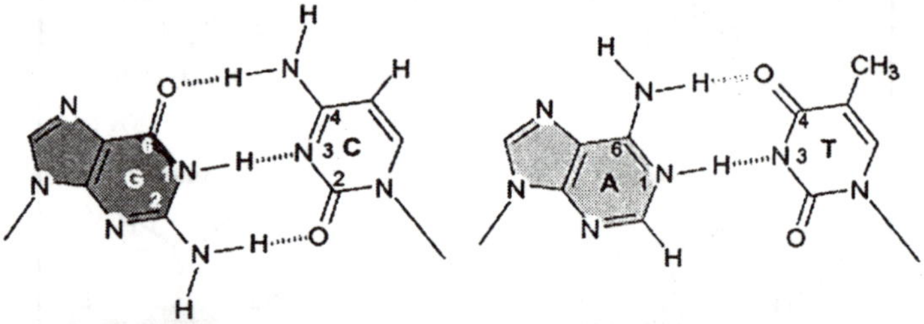

Fig. The Figure shows Hydrogen Bonding between the Bases

A corresponding arrangement can be drawn between a guanine and a cytosine, so that there is both hydrogen bonding and shape complementarity in this base pair as well. A G:C base pair has three hydrogen bonds, because the exocyclic NH_2 at C_2 on guanine lies opposite to, and can hydrogen bond with, a carbonyl at C_2 on cytosine. Likewise, a hydrogen bond can form between N_1 of guanine and N_3 of cytosine and between the carbonyl at C_6 of guanine and the exocyclic NH_2 at C_4 of cytosine.

Watson-Crick base pairing requires that the bases are in their preferred tautomeric states. An important feature of the double helix is that the two base pairs have exactly the same geometry; having an A:T base pair or a G:C base pair between the two sugars does not perturb the arrangement of the sugars. Neither does T:A or C:G. In other words, there is an approximately twofold axis of symmetry that relates the two sugars and all four base pairs can be accommodated within the same arrangement without any distortion of the overall structure of the DNA.

HYDROGEN BONDING

The hydrogen bonds between complementary bases are a fundamental feature of the double helix, contributing to the thermodynamic stability of the helix and providing the information content and specificity of base pairing. Hydrogen bonding might not at first glance appear to contribute importantly to the stability of DNA for the following reason. An organic molecule in aqueous solution has all of its hydrogen bonding properties satisfied by water molecules that come on and off very rapidly. As a result, for every hydrogen bond that is made when a base pair forms, a hydrogen bond with water is broken that was there before the base pair formed.

Thus, the net energetic contribution of hydrogen bonds to the stability of the double helix would appear to be modest. However, when polynucleotide strands are separate, water molecules are lined up on the bases. When strands come together in the double helix, the water molecules are displaced from the bases. This creates disorder and increases entropy, thereby stabilizing the double helix. Hydrogen bonds are not the only force that stabilizes the double helix. A second important contribution comes from stacking interactions between the bases. The bases are flat, relatively waterinsoluble molecules, and they tend to stack above each other roughly perpendicular to the direction of the helical axis.

Electron cloud interactions between bases in the helical stacks contribute significantly to the stability of the double helix. Hydrogen bonding is also important for the specificity of base pairing. Suppose we tried to pair an adenine with a cytosine. Then we would have a hydrogen bond acceptor (N_1 of adenine) lying opposite a hydrogen bond acceptor (N_3 of cytosine) with no room to put a water molecule in between to satisfy the two acceptors. Likewise, two hydrogen bond donors, the NH_2 groups at C_6 of adenine and C_4 of cytosine, would lie opposite each other. Thus, an A:C base pair would be unstable because water would have to be stripped off the donor and acceptor groups without restoring the hydrogen bond formed within the base pair.

BASES CAN FLIP OUT FROM THE DOUBLE HELIX

As we have seen, the energetics of the double helix favour the pairing of each base on one polynucleotide strand with the complementary base on the other strand. Certain enzymes that methylate bases or remove damaged bases do so with the base in an extra helical configuration in which it is flipped out from the double helix, enabling the base to sit in the catalytic cavity of the enzyme.

Furthermore, enzymes involved in homologous recombination and DNA repair are believed to scan DNA for homology or lesions by flipping out one base after another. This is not energetically expensive because only one base is flipped out at a time. Clearly, DNA is more flexible than might be assumed at first glance.

DNA IS USUALLY A RIGHT-HANDED DOUBLE HELIX

Applying the handedness rule from physics, we can see that each of the polynucleotide chains in the double helix is right-handed. In our mind's eye, hold our right hand up to the DNA molecule in Figure with our thumb pointing up and along the long axis of the helix and our fingers following the grooves in the helix.

Trace along one strand of the helix in the direction in which our thumb is pointing. Notice that we go around the helix in the same direction as our fingers are pointing. This does not work if we use our left hand. Try it! A consequence of the helical nature of DNA is its periodicity.

Each base pair is displaced (twisted) from the previous one by about 36°. Thus, in the X-ray crystal structure of DNA it takes a stack of about 10 base pairs to go completely around the helix (360°). That is, the helical periodicity is generally 10 base pairs per turn of the helix.

DOUBLE HELIX

As a result of the double-helical structure of the two chains, the DNA molecule is a long extended polymer with two grooves that are not equal in size to each other. Why are there a minor groove and a major groove? It is a simple consequence of the geometry of the base pair. The angle at which the two sugars protrude from the base pairs (that is, the angle between the glycosidic bonds) is about 120° (for the narrow angle or 240° for the wide angle).

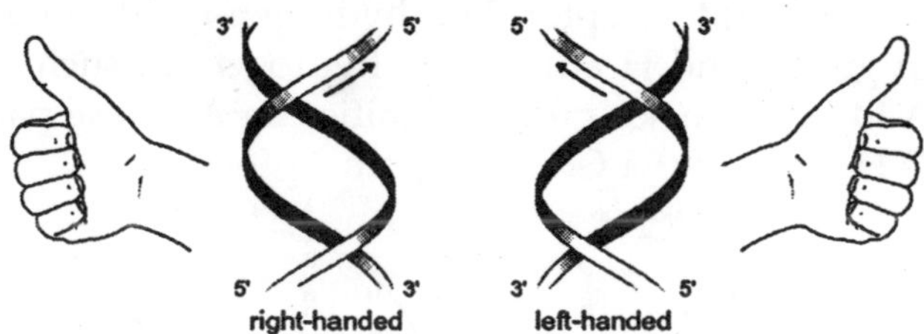

Fig. Left- and Right-Handed Helices

As a result, as more and more base pairs stack on top of each other, the narrow angle between the sugars on one edge of the base pairs generates a minor groove and the large angle on the other edge generates a major groove. (If the sugars pointed away from each other in a straight line, that is, at an angle of 180°, then two grooves would be of equal dimensions and there would be no minor and major grooves.)

CHEMICAL INFORMATION OF DNA

The edges of each base pair are exposed in the major and minor grooves, creating a pattern of hydrogen bond donors and acceptors and of van der Waals surfaces that identifies the base pair. The edge of an A:T base pair displays the following chemical groups in the following order in the major groove: a hydrogen bond acceptor (the N_7 of adenine), a hydrogen bond donor (the exocyclic amino group on C_6 of adenine), a hydrogen bond acceptor (the carbonyl group on C_4 of thymine) and a bulky hydrophobic surface (the methyl group on C_5 of thymine).

Similarly, the edge of a G:C base pair displays the following groups in the major groove: a hydrogen bond acceptor (at N_7 of guanine), a hydrogen bond acceptor (the carbonyl on C_6 of guanine), a hydrogen bond donor (the exocyclic amino group on C_4 of cytosine), a small non-polar hydrogen (the hydrogen at C_5 of cytosine).

Thus, there are characteristic patterns of hydrogen bonding and of overall shape that are exposed in the major groove that distinguish an A:T base pair from a G:C base pair, and, for that matter, A:T from T:A, and G:C from C:G. We can think of these features as a code in which A represents a hydrogen bond acceptor, D a hydrogen bond donor, M a methyl group, and H a nonpolar hydrogen. In such a code, A D A M in the major groove signifies an A:T base pair, and A A D H stands for a G:C base pair.

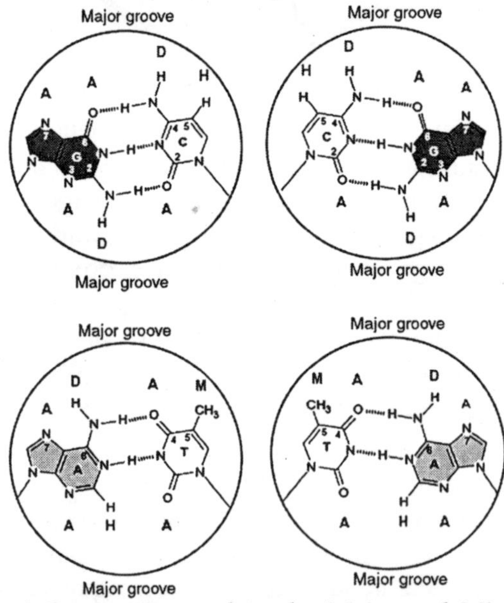

Fig. Chemical Groups Exposed in the Major and Minor Grooves from the Edges of the Base Pairs. The Letters Identify as Hydrogen Bond Acceptors (A), Hydrogen Bond Donors (D), Nonpolar Hydrogens (H), and Methyl groups (M)

Likewise, M A D A stands for a T:A base pair and H D A A is characteristic of a C:G base pair. In all cases, this code of chemical groups in the major groove specifies the identity of the base pair. These patterns are important because they allow proteins to unambiguously recognize DNA sequences without having to open and thereby disrupt the double helix. Indeed, as we shall see, a principal decoding mechanism relies upon the ability of amino acid side chains to protrude into the major groove and to recognize and bind to specific DNA sequences. The minor groove is not as rich in chemical information and what information is available is less useful for distinguishing between base pairs. The small size of the minor groove is less able to accommodate amino acid side chains. Also, A:T and T:A base pairs and G:C and C:G pairs look similar to one another in the minor groove.

An A:T base pair has a hydrogen bond acceptor (at N_3 of adenine), a nonpolar hydrogen (at N_2 of adenine) and a hydrogen bond acceptor (the carbonyl on C_2 of thymine). Thus, its code is A H A. But this code is the same if read in the opposite direction, and hence an A:T base pair does not look very different from a T:A base pair from the point of view of the hydrogen bonding properties of a protein poking its side chains into the minor groove.

Likewise, a G:C base pair exhibits a hydrogen bond acceptor (at N_3 of guanine), a hydrogen bond donor (the exocyclic amino group on C_2 of guanine), and a hydrogen bond acceptor (the carbonyl on C_2 of cytosine), representing the code A D A. Thus, from the point of view of hydrogen bonding, C:G and G:C base pairs do not look very different from each other either. The minor groove does look different when comparing an A:T base pair with a G:C base pair, but G:C and C:G, or A:T and T:A, cannot be easily distinguished.

DOUBLE HELIX IN MULTIPLE CONFORMATIONS

Early X-ray diffraction studies of DNA, which were carried out using concentrated solutions of DNA that had been drawn out into thin fibers, revealed two kinds of structures, the B and the A forms of DNA. The B form, which is observed at high humidity, most closely corresponds to the average structure of DNA under physiological conditions. It has 10 base pairs per turn, and a wide major groove and a narrow minor groove. The A form, which is observed under conditions of low humidity, has 11 base pairs per turn. Its major groove is narrower and much deeper than that of the B form, and its minor groove is broader and shallower. The vast majority of the DNA in the cell is in the B form, but DNA does adopt the

A structure in certain DNA-protein complexes. Also, as we shall see, the A form is similar to the structure that RNA adopts when double helical.

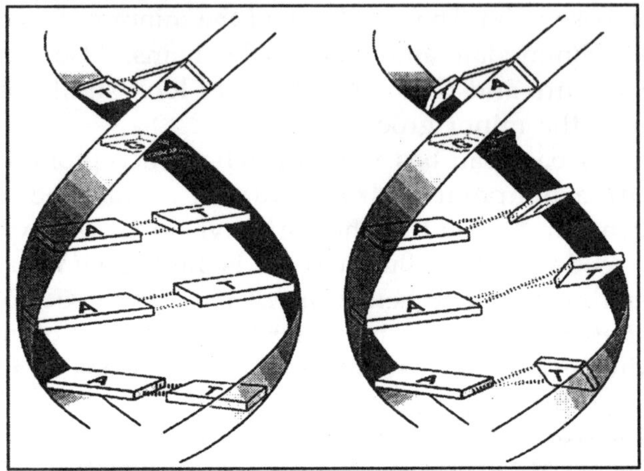

Fig. The Propeller Twist between the Purine and Pyrimidine Base Pairs of a Right-Handed Helix

The B form of DNA represents an ideal structure that deviates in two respects from the DNA in cells. First, DNA in solution, as we have seen, is somewhat more twisted on average than the B form, having on average 10.5 base pairs per turn of the helix. Second, the B form is an average structure whereas real DNA is not perfectly regular. Rather, it exhibits variations in its precise structure from base pair to base pair. This was revealed by comparison of the crystal structures of individual DNAs of different sequences. For example, the two members of each base pair do not always lie exactly in the same plane.

Rather, they can display a "propeller twist" arrangement in which the two flat bases counter rotate relative to each other along the long axis of the base pair, giving the base pair a propeller-like character. Moreover, the precise rotation per base pair is not a constant. As a result, the width of the major and minor grooves varies locally.

Thus, DNA molecules are never perfectly regular double helices. Instead, their exact conformation depends on which base pair (A:T, T:A, G:C, or C:G) is present at each position along the double helix and on the identity of neighboring base pairs. Still, the B form is for many purposes a good first approximation of the structure of DNA in cells.

DNA CAN FORM A LEFT-HANDED HELIX

DNA containing alternative purine and pyrimidine residues can fold into left-handed as well as right-handed helices. To understand how DNA can form a left-handed helix, we need to consider the glycosidic bond that connects the base to the 1′ position of 2′-deoxyribose. This bond can be in one of two conformations called *syn* and *anti*. In right-handed DNA, the glycosidic bond is always in the *anti* conformation. In the left-handed helix, the fundamental repeating unit usually is a purine-pyrimidine dinucleotide, with the glycosidic bond in the *anti* conformation at pyrimidine residues and in the *syn* conformation at purine residues. It is this *syn* conformation at the purine nucleotides that is responsible for the left-handedness of the helix.

The change to the *syn* position in the purine residues to alternating *anti–syn* conformations gives the backbone of left-handed DNA a zigzag look, which distinguishes it from right-handed forms. The rotation that effects the change from *anti* to *syn* also causes the ribose group to undergo a change in its pucker. In solution alternating purine–pyrimidine residues assume the left-handed conformation only in the presence of high concentrations of positively charged ions (e.g., Na′) that shield the negatively charged phosphate groups. At lower salt concentrations, they form typical right-handed conformations. The physiological significance of Z DNA is uncertain and left-handed helices probably account at most for only a small of proportion of a cell's DNA.

DNA STRANDS CAN SEPARATE

Because the two strands of the double helix are held together by relatively weak (non-covalent) forces, we might expect that the two strands could come apart easily. Indeed, the original structure for the double helix suggested that DNA replication would occur in just this manner. The complementary strands of double helix can also be made to come apart when a solution of DNA is heated above physiological temperatures (to near 100°C) or under conditions of high pH, a process known as denaturation.

However, this complete separation of DNA strands by denaturation is reversible. When heated solutions of denatured DNA are slowly cooled, single strands often meet their complementary strands and reform regular double helices. The capacity to renature denatured DNA molecules permits artificial hybrid DNA molecules to be formed by slowly cooling mixtures of denatured DNA from two different sources. Likewise, hybrids can be formed between complementary strands of DNA and RNA.

The ability to form hybrids between two single-stranded nucleic acids (hybridization) is the basis for several indispensable techniques in molecular biology, such as Southern blot hybridization and DNA microarrays. Important insights into the properties of the double helix were obtained from classic experiments carried out in the 1950s in which the denaturation of DNA was studied under a variety of conditions.

In these experiments DNA denaturation was monitored by measuring the absorbance of ultraviolet light passed through a solution of DNA. DNA maximally absorbs ultraviolet light at a wavelength of about 260 nm. It is the bases that are principally responsible for this absorption. When the temperature of a solution of DNA is raised to near the boiling point of water, the optical density (absorbance) at 260 nm markedly increases.

The explanation for this increase is that duplex DNA is hypochromic; it absorbs less ultraviolet light by about 40% than do individual DNA chains. The hypochromicity is due to base stacking, which diminishes the capacity of the bases in duplex DNA to absorb ultraviolet light.

If we plot the optical density of DNA as a function of temperature, we observe that the increase in absorption occurs abruptly over a relatively narrow temperature range. The midpoint of this transition is the melting point or Tm. Like ice, DNA melts: it undergoes a transition from a highly ordered double-helical structure to a much less ordered structure of individual strands.

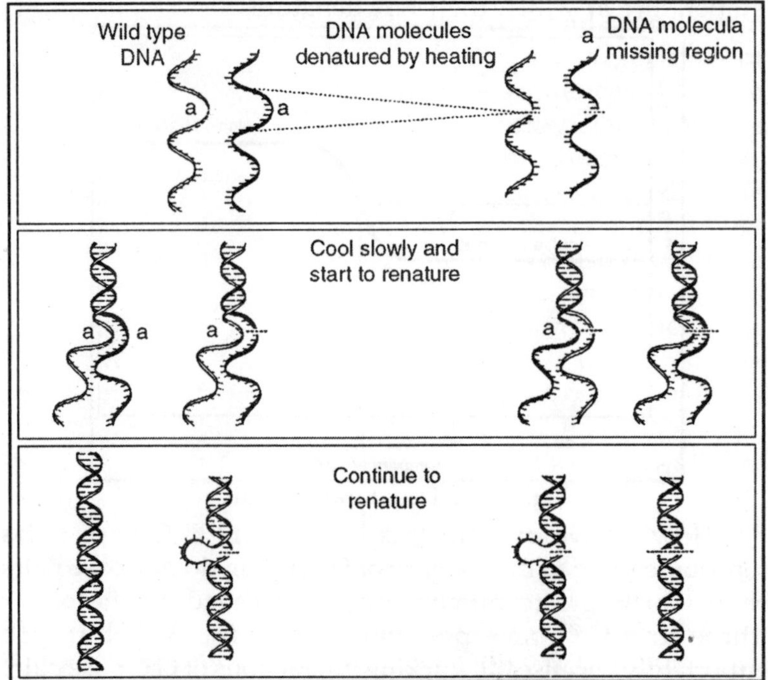

Fig. Reannealing and Hybridization

The sharpness of the increase in absorbance at the melting temperature tells us that the denaturation and renaturation of complementary DNA strands is a highly cooperative, zippering-like process.

Renaturation, for example, probably occurs by means of a slow nucleation process in which a relatively small stretch of bases on one strand find and pair with their complement on the complementary strand. The remainder of the two strands then rapidly zipper-up from the nucleation site to reform an extended double helix.

The melting temperature of DNA is a characteristic of each DNA that is largely determined by the G:C content of the DNA and the ionic strength of the solution. The higher the percent of G:C base pairs in the DNA (and hence the lower the content of A:T base pairs), the higher the melting point. Likewise, the higher the salt concentration of the solution the greater the temperature at which the DNA denatures.

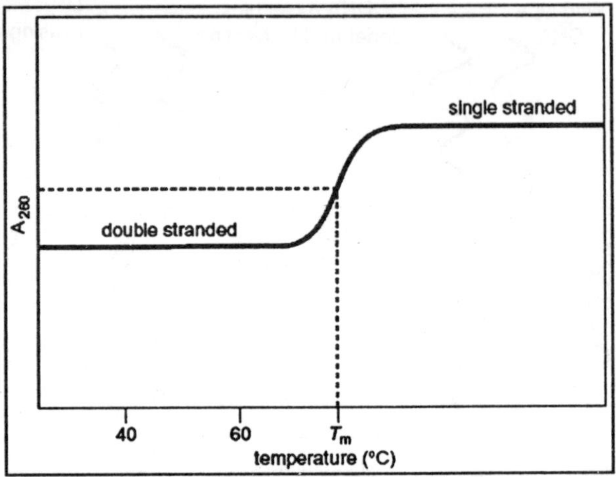

Fig. DNA Denaturation Curve

How do we explain this behaviour? G:C base pairs contribute more to the stability of DNA than do A:T base pairs because of the greater number of hydrogen bonds for the former (three in a G:C base pair versus two for A:T) but also importantly because the stacking interactions of G:C base pairs with adjacent base pairs are more favorable than the corresponding interactions of A:T base pairs with their neighboring base pairs.

The effect of ionic strength reflects another fundamental feature of the double helix. The backbones of the two DNA strands contain phosphoryl groups, which carry a negative charge. These negative charges are close enough across the two strands that if not shielded they tend to cause the strands to repel each other, facilitating their separation. At high ionic strength, the negative charges are shielded by cations, thereby stabilizing the helix. Conversely, at low ionic strength the unshielded negative charges render the helix less stable.

CIRCLES CIRCULAR DNA MOLECULES

It was initially believed that all DNA molecules are linear

and have two free ends. Indeed, the chromosomes of eukaryotic cells each contain a single (extremely long) DNA molecule. But now we know that some DNAs are circles. For example, the chromosome of the small monkey DNA virus SV40 is a circular, double-helical DNA molecule of about 5,000 base pairs. Also, most (but not all) bacterial chromosomes are circular; *E. coli* has a circular chromosome of about 5 million base pairs.

Additionally, many bacteria have small autonomously replicating genetic elements known as plasmids, which are generally circular DNA molecules. Interestingly, some DNA molecules are sometimes linear and sometimes circular. The most well-known example is that of the bacteriophage ', a DNA virus of *E. coli*. The phage ' genome is a linear double-stranded molecule in the virion particle. However, when the genome is injected into an *E. coli* cell during infection, the DNA circularizes. This occurs by base-pairing between single-stranded regions that protrude from the ends of the DNA and that have complementary sequences ("sticky ends").

DNA TOPOLOGY

As DNA is a flexible structure, its exact molecular parameters are a function of both the surrounding ionic environment and the nature of the DNA-binding proteins with which it is complexed. Because their ends are free, linear DNA molecules can freely rotate to accommodate changes in the number of times the two chains of the double helix twist about each other. But if the two ends are covalently linked to form a circular DNA molecule and if there are no interruptions in the sugar phosphate backbones of the two strands, then the absolute number of times the chains can twist about each other cannot change. Such a covalently closed, circular DNA is said to be topologically constrained.

Even the linear DNA molecules of eukaryotic chromosomes are subject to topological constraints due to their entrainment in chromatin and interaction with other cellular components. Despite these constraints, DNA participates in numerous dynamic processes in the cell. For example, the two strands of the double helix, which are twisted around each other, must rapidly separate in order for

DNA to be duplicated and to be transcribed into RNA. Thus, understanding the topology of DNA and how the cell both accommodates and exploits topological constraints during DNA replication, transcription, and other chromosomal transactions is of fundamental importance in molecular biology.

CIRCULAR DNA

Let us consider the topological properties of covalently closed, circular DNA, which is referred to as cccDNA. Because there are no interruptions in either polynucleotide chain, the two strands of cccDNA cannot be separated from each other without the breaking of a covalent bond. If we wished to separate the two circular strands without permanently breaking any bonds in the sugar phosphate backbones,we would have to pass one strand through the other strand repeatedly(we will encounter an enzyme that can perform just this feat!).

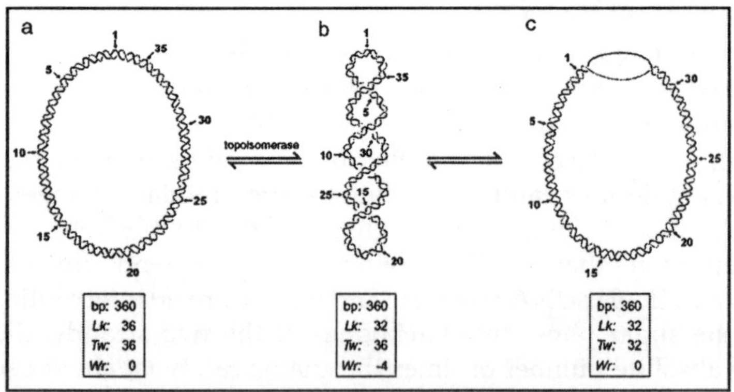

Fig. Topological States of Covalently Closed Circular (ccc) DNA

The number of times one strand would have to be passed through the other strand in order for the two strands to be entirely separated from each other is called the linking number. The linking number, which is always an integer, is an invariant topological property of cccDNA, no matter how much the DNA molecule is distorted.

Linking Number Is Composed of Twist and Writhe

The linking number is the sum of two geometric components called the twist and the writhe. Let us consider twist first. Twist is simply the number of helical turns of one strand about the other, that is, the number of times one strand completely wraps around the other strand. Consider a cccDNA that is lying flat on a plane. In this flat conformation, the linking number is fully composed of twist. Indeed, the twist can be easily determined by counting the number of times the two strands cross each other. The helical crossovers (twist) in a right-handed helix are defined as positive such that the linking number of DNA will have a positive value. But cccDNA is generally not lying flat on a plane.

Rather, it is usually torsionally stressed such that the long axis of the double helix crosses over itself, often repeatedly, in three-dimensional space. This is called *writhe.* To visualize the distortions caused by torsional stress, think of the coiling of a telephone cord that has been overtwisted. Writhe can take two forms.

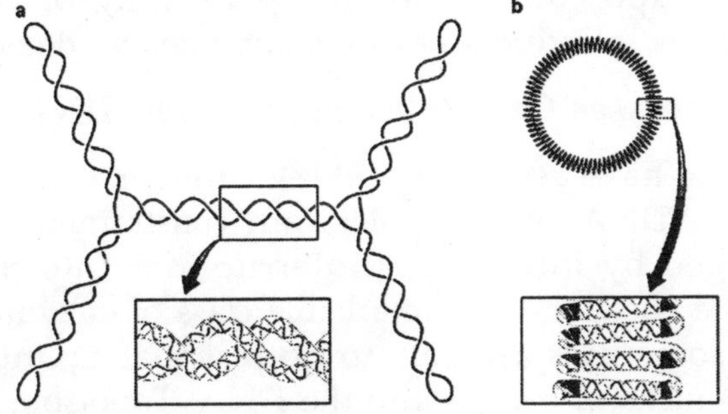

Fig. Two Forms of Writhe of Supercoiled DNA

One form is the interwound or plectonemic writhe, in which the long axis is twisted around itself. The other form of writhe is a toroid or spiral in which the long axis is wound in a cylindrical manner, as often occurs when DNA wraps around protein. The writhing number (*Wr*) is the total number of interwound and/or spiral writhes in cccDNA. For example, the molecule has a writhe of 4 from 4 interwound writhes. Interwound writhe and spiral writhe are topologically equivalent to each other and are readily interconvertible geometric properties of cccDNA.

Also, twist and writhe are interconvertible. A molecule of cccDNA can readily undergo distortions that convert some of its twist to writhe or some of its writhe to twist without the breakage of any covalent bonds.

The only constraint is that the sum of the twist number (*Tw*) and the writhing number (*Wr*) must remain equal to the linking number (*Lk*). This constraint is described by the equation: $Lk = Tw + Wr$.

Negative Supercoiling in Eukaryotes

DNA in the nucleus of eukaryotic cells is packaged in small particles known as nucleosomes in which the double helix is wrapped almost two times around the outside circumference of a protein core. We will be able to recognize this wrapping as the toroid or spiral form of writhe. Importantly, it occurs in a lefthanded manner.. It turns out that writhe in the form of left-handed spirals is equivalent to negative supercoils. Thus, the packaging of DNA into nucleosomes introduces negative superhelical density.

Topoisomerases Can Relax Supercoiled DNA

As we have seen, the linking number is an invariant property of DNA that is topologically constrained. It can only be changed by introducing interruptions into the sugar-phosphate backbone. A remarkable class of enzymes known as topoisomerases are able to do just that by introducing transient nicks or breaks into the DNA. Topoisomerases are of two broad types. Type II topoisomerases change the linking number in steps of two. They make transient double-stranded breaks in the DNA, through which they pass a region of uncut duplex DNA before resealing the break. Type II topoisomerases require energy from ATP hydrolysis for their action. Type I topoisomerases, in contrast, change the linking number of DNA in steps of one. They make transient singlestranded breaks in the DNA, allowing one strand to pass through the break in the other before resealing the nick.

Type I topoisomerases relax DNA by removing supercoils (dissipating writhe).

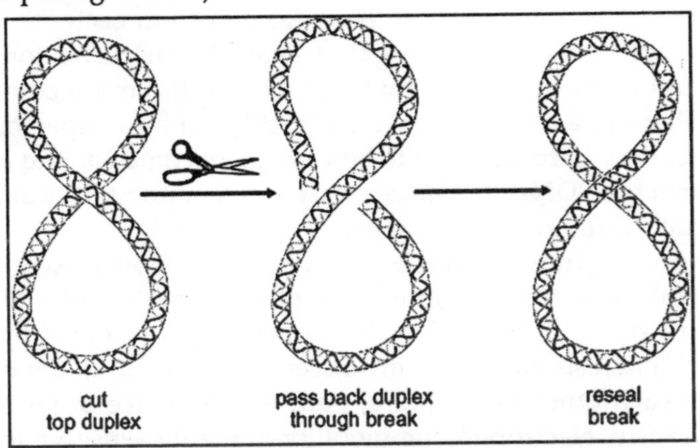

Fig. Schematic for Changing the Linking Number in DNA with Topoisomerase II

They can be compared to the protocol of introducing nicks into cccDNA with DNase and then repairing the nicks, which as we saw can be used to relax cccDNA, except that type I topoisomerases relax DNA in a controlled and concerted manner. In contrast to type II topoisomerases, type I topoisomerases do not require ATP. Both type I and type II topoisomerases work through an intermediate in which the enzyme is covalently attached to one end of the broken DNA.

SPECIAL TOPOISOMERASE

Both prokaryotes and eukaryotes have type I and type II topoisomerases, which are capable of removing supercoils from DNA. In addition, however, prokaryotes have a special type II topoisomerase known as DNA gyrase that is able to introduce negative supercoils, rather than remove them.

DNA gyrase is responsible for the negative supercoiling of chromosomes in prokaryotes, which facilitates unwinding of the DNA duplex during transcription and DNA replication.

DNA Topoisomers

Covalently closed, circular DNA molecules of the same

length but of different linking numbers are called DNA topoisomers. Even though topoisomers have the same molecular weight, they can be separated from each other by electrophoresis through a gel of agarose. The basis for this separation is that the greater the writhe the more compact the shape of a cccDNA. Once again, think of how supercoiling a telephone cord causes it to become more compact. The more compact the DNA, the more easily (up to a point) it is able to migrate through the gel matrix.

Thus, a fully relaxed cccDNA migrates more slowly than a highly supercoiled topoisomer of the same circular DNA. Figure shows a ladder of DNA topoisomers resolved by gel electrophoresis. Molecules in adjacent rungs of the ladder differ from each other by a linking number difference of just one. Obviously, electrophoretic mobility is highly sensitive to the topological state of DNA.

DNA Has a Helical Periodicity

The observation that DNA topoisomers can be separated from each other electrophoretically is the basis for a simple experiment that proves that DNA has a helical periodicity of about 10.5 base pairs per turn in solution. Consider three cccDNAs of sizes 3990, 3995, and 4011 base pairs that were relaxed to completion by treatment with topoisomerase I. When subjected to electrophoresis through agarose, the 3990- and 4011-base-pair DNAs exhibit essentially identical mobilities.

Due to thermal fluctuation, topoisomerase treatment actually generates a narrow spectrum of topoisomers, but for simplicity let us consider the mobility of only the most abundant topoisomer (that corresponding to the cccDNA in its most relaxed state). The mobilities of the most abundant topoisomers for the 3990- and 4011-base-pair DNAs are indistinguishable because the 21-base-pair difference between them is negligible compared to the sizes of the rings.

The most abundant topoisomer for the 3995-base-pair ring, however, is found to migrate slightly more rapidly than the other two rings even though it is only 5 base pairs larger than the 3990-base-pair ring. How are we to explain this anomaly? The 3990- and 4011- base-pair rings in their most relaxed states are expected to have linking numbers equal to Lk^O, that is, 380 in the case of the 3990-base-pair ring (dividing the size by 10.5 base pairs) and 382 in the case of the 4011-base-pair ring. Because *Lk* is equal to Lk^O, the linking difference ('$Lk = Lk - Lk^O$) in both cases is zero and there is no writhe.

But because the linking number must be an integer, the most relaxed state for the 3995-base-pair ring would be either of two topoisomers having linking numbers of 380 or 381. However, Lk^O for the 3995-base-pair ring is 380.5. Thus, even in its most relaxed state, a covalently closed circle of 3995 base pairs would necessarily have about half a unit of writhe (its linking difference would be 0.5), and hence it would migrate more rapidly than the 3990- and 4011-base-pair circles.

In other words, to explain how rings that differ in length by 21 base pairs (two turns of the helix) have the same mobility whereas a ring that differs in length by only 5 base pairs (about half a helical turn) exhibits a different mobility, we must conclude that DNA in solution has a helical periodicity of about 10.5 base pairs per turn.

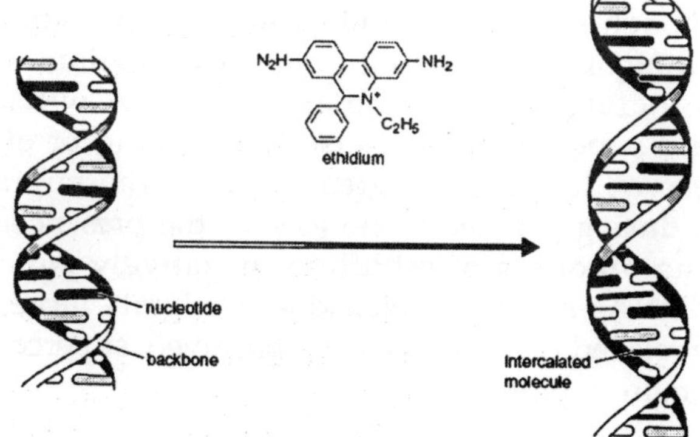

Fig. Intercalation of Ethidium Bromide into DNA

Ethidium Ions Cause DNA to Unwind

Ethidium is a large, flat, multi-ringed cation. Its planar shape enables ethidium to slip (intercalate) between the stacked base pairs of DNA. Because it fluoresces when exposed to ultraviolet light, and because its fluorescence increases dramatically after intercalation, ethidium is used as a stain to visualize DNA. When an ethidium ion intercalates between two base pairs, it causes the DNA to unwind by 26°, reducing the normal rotation per base pair from ~36° to ~10°. In other words, ethidium decreases the twist of DNA.

Imagine the extreme case of a DNA molecule that has an ethidium ion between every base pair. Instead of 10 base pairs per turn it would have 36! When ethidium binds to linear DNA or to a nicked circle, it simply causes the helical pitch to increase. But consider what happens when ethidium binds to covalently closed, circular DNA.

The linking number of the cccDNA does not change (no covalent bonds are broken and resealed), but the twist decreases by 26° for each molecule of ethidium that has bound to the DNA. Because $Lk = Tw + Wr$, this decrease in Tw must be compensated for by a corresponding increase in Wr. If the circular DNA is initially negatively supercoiled (as is normally the case for circular DNAs isolated from cells), then the addition of ethidium will increase Wr. In other words, the addition of ethidium will relax the DNA.

If enough ethidium is added, the negative supercoiling will be brought to zero, and if even more ethidium is added, Wr will increase above zero and the DNA will become positively supercoiled. Because the binding of ethidium increases Wr, its presence greatly affects the migration of cccDNA during gel electrophoresis. In the presence of non-saturating amounts of ethidium, negatively supercoiled circular DNAs are more relaxed and migrate more slowly, whereas relaxed cccDNAs become positively supercoiled and migrate more rapidly.

Chapter 2

Principles of Genetics

Genetics, study of the function and behaviour of genes. Genes are bits of biochemical instructions found inside the cells of every organism from bacteria to humans. Offspring receive a mixture of genetic information from both parents. This process contributes to the great variation of traits that we see in nature, such as the colour of a flower's petals, the markings on a butterfly's wings, or such human behavioral traits as personality or musical talent.

Geneticists seek to understand how the information encoded in genes is used and controlled by cells and how it is transmitted from one generation to the next. Geneticists also study how tiny variations in genes can disrupt an organism's development or cause disease. Increasingly, modern genetics involves genetic engineering, a technique used by scientists to manipulate genes. Genetic engineering has produced many advances in medicine and industry, but the potential for abuse of this technique has also presented society with many ethical and legal controversies.

Genetic information is encoded and transmitted from generation to generation in deoxyribonucleic acid (DNA). DNA is a coiled molecule organized into structures called chromosomes within cells. Segments along the length of a DNA molecule form genes. Genes direct the synthesis of proteins, the molecular laborers that carry out all life-supporting activities in the cell. Although all humans share the same set of genes, individuals can inherit different forms of a given gene, making each person genetically unique.

Since the earliest days of plant and animal domestication,

around 10,000 years ago, humans have understood that characteristic traits of parents could be transmitted to their offspring. The first to speculate about how this process worked were Greek scholars around the 4th century BC, who promoted theories based on conjecture or superstition.

Some of these theories remained in favour for several centuries. The scientific study of genetics did not begin until the late 19th century. In experiments with garden peas, Austrian monk Gregor Mendel described the patterns of inheritance, observing that traits were inherited as separate units. These units are now known as genes. Mendel's work formed the foundation for later scientific achievements that heralded the era of modern genetics.

The Importance of Genetics

The modern science of genetics influences many aspects of daily life, from the food we eat to how we identify criminals or treat diseases. In agriculture, genetic advances enable scientists to alter a plant or animal to make it more useful. For instance, some food crops, such as oranges, potatoes, wheat, and rice, have been genetically altered to withstand insect pests, resulting in a higher crop yield. Tomatoes and apples have been modified so that they resist discoloration or bruising on their way to market, enhancing their appeal on supermarket shelves.

The genetic makeup of cows has been modified to increase their milk production, and cattle raised for beef have been altered so that they grow faster. Genetic technologies have also helped convict criminals. DNA recovered from semen, blood, skin cells, or hair found at a crime scene can be analyzed in a laboratory and compared with the DNA of a suspect. An individual's DNA is as unique as a set of fingerprints, and a DNA match can be used in a courtroom as evidence connecting a person to a crime.

Genetics has revolutionized the way industries produce certain substances, many of which formerly required costly and arduous manufacturing methods. In medicine, scientists can genetically alter bacteria so that they mass-produce specific proteins, such as insulin used by people with diabetes mellitus or human growth hormone used by children who suffer from growth disorders. In other medical applications, genetic technologies have been instrumental in the development of gene therapy. In this still-experimental form of treatment, scientists try to cure disease by replacing malfunctioning genes with healthy ones.

Gene therapy has shown promise in treating some devastating conditions, including some forms of cancer and cystic fibrosis. Genetically engineered vaccines are being tested for possible use against the human immunodeficiency virus (HIV), the virus that causes acquired immunodeficiency syndrome (AIDS). The field of human genetics has been energized in recent years by the Human Genome Project, an international collaboration of scientists, governments, and drug companies from around the world.

Scientists working on this project have developed detailed maps that identify the chromosomal locations of the estimated 20,000 to 25,000 human genes. The vast databases emerging from the project help scientists study previously unknown genes as well as many genes all at once to examine how gene activity can cause disease. Scientists expect that the project will lead to the development of new drugs targeted to specific genetic disorders.

Despite the benefits derived from genetic advancements, some observers have voiced concerns that genetically engineered organisms could harm people or the environment. Others fear that new genetic technologies may enable scientists to modify genes that affect characteristics other than those responsible for disease.

They warn that determining who has undesirable genetic characteristics may lead to discriminatory practices. Others are concerned about the common misperception that a person's genes determine all aspects of a person's life, including health and behaviour. This misperception leads people to blame their genetic makeup for problems, leaving no room for the influence of free will, personal responsibility, or hope for change. These and other challenging issues place geneticists at the crossroads of science and social responsibility, where they work to promote understanding of genetic advances and prevent the abuse of them.

PRINCIPLES OF GENETICS

The site where genes work is the cell. Some organisms, such as paramecia or amoebas, are made up of a single cell. Other organisms are made of many kinds of cells, each having a different function. For instance, a tree contains some cells that form the root system and other cells that form leaves. Each cell's function within an organism is determined by the genetic information encoded in DNA.

In animals, plants, and other eukaryotes (organisms whose cells contain a nucleus), DNA resides within membrane-bound structures in the cell. These structures include the nucleus, the energy-producing mitochondria, and, in plants, the chloroplasts (structures where photosynthesis takes place). In prokaryotes, one-celled organisms and bacteria that lack internal membrane-bound structures, DNA floats freely within the cell body.

Cell Division and Reproduction

Organisms could not grow or function properly if the genetic information encoded in DNA was not passed from cell to cell. DNA is packaged into structures called chromosomes within a cell. Every chromosome in a cell contains many genes, and each gene is located at a particular site, or locus, on the chromosome. Chromosomes vary in size and shape and usually occur in matched pairs called homologues. The number of homologous chromosomes in a cell depends upon the organism—for example, most cells in the human body contain 23 pairs of chromosomes, while most cells of the fruit fly *Drosophila* contain 4 pairs.

Within all organisms, cells divide to produce new cells, each of which requires the genetic information found in DNA. Yet simply splitting the DNA of a dividing cell between two new cells would lead to disaster—the two new cells would have different instructions and each subsequent generation of cells would have less and less genetic information to work with. Imagine how chaotic it would be to rip an architectural blueprint in two, give each half to different contractors, and tell them to construct identical buildings. Just as each contractor would require a full copy of the blueprint to construct a complete building, each new cell needs a complete copy of an organism's genetic information to function properly.

Organisms use two types of cell division to ensure that DNA is passed down from cell to cell during reproduction. Simple one-celled organisms and other organisms that reproduce asexually—that is, without the joining of cells from two different organisms—reproduce by a process called mitosis. During mitosis a cell doubles its DNA before dividing into two cells and distributing the DNA evenly to each resulting cell. Organisms that reproduce sexually use a different type of cell division. These organisms produce special cells called gametes, or egg and sperm. In the cell division known as meiosis, the chromosomes in a gamete cell are reduced by half. During sexual reproduction, an egg and sperm unite to form a zygote, in which the full number of chromosomes is restored.

Mitosis

Mitosis occurs in five stages: prophase, prometaphase, metaphase, anaphase, and telophase. During prophase, the start of mitosis, the DNA of each chromosome replicates. Each chromosome then reorganizes into paired structures called sister chromatids, with each member of the pair containing a full copy of the DNA sequence.

The sister chromatids condense, thickening until they appear joined at a single site, known as the centromere. Prometaphase is marked by the disintegration of the nuclear membrane. The sister chromatids line up in the middle of the cell halfway between the poles. In metaphase, exactly half of the chromatids face one pole, and half face the other pole. This equilibrium position is called the metaphase plate. In anaphase, the chromatid pairs split apart at the centromere, and each half of the pair then moves toward opposite poles of the cells. In telophase, the final stage of mitosis, a nuclear membrane forms around the chromosomes at each pole of the cell. Mitosis ends with the formation of two new cells, each with a matching full set of chromosomes as well as an identical complement of cellular structures.

Meiosis

During meiosis, two cell divisions occur to produce four daughter cells from the original parent cell. Each resulting cell has half the chromosomal DNA of the parent cell. A half set of chromosomes in an organism is known as the haploid number. In the first cell division of meiosis the chromosomes of a gamete cell duplicate and join in pairs. The paired chromosomes align at the equator of the cell, and then separate and move to opposite poles in the cell. The cell then splits to form two daughter cells. As meiosis proceeds, the two daughter cells undergo another cell division to form four cells, each of which bears half of the number of chromosomes found in the other cells of the organism.

Meiosis ensures that reproduction will produce a zygote that has received one set of chromosomes from the male parent and one set of chromosomes from the female parent to form a full set of chromosomes. The entire set of chromosomes in an organism is known as the diploid number. Once formed, the zygote continues to divide and grow through the process of mitosis.

Patterns of Inheritance

In life forms that reproduce asexually, such as bacteria and amoebas, all offspring share the exact same genes and are identical to their parents. The genetic transmission that occurs in organisms that reproduce sexually is far more complex. An individual that forms by the union of two gametes inherits its chromosomes from two distinct parents. Consequently, sexual reproduction guarantees that offspring with new combinations of genes will continually arise.

Certain patterns of inheritance were evident long before

scientists discovered the molecular structure of DNA and chromosomes. Throughout history, people have recognized that certain traits, whether in humans, animals, or agricultural crops, could be passed from generation to generation. Yet for centuries, people were unable to reconcile many confusing observations about the mechanisms of inheritance.

The first person to make sense of this complex subject was Austrian monk Gregor Mendel, who conducted a series of experiments on pea plants beginning in the 1850s. Mendel observed the results of crossbreeding plants with different characteristics, such as height, flower colour, and seed shape. His conclusions from these experiments led him to develop explanations for how traits are transmitted from generation to generation. Mendel's theories form the foundation of modern genetics.

Mendel's Rules

In his research, Mendel observed that characteristics were inherited as separate units, each of which was inherited independently of the others. Mendel suggested that each parent has pairs of these units but contributes only one of each pair to offspring. The units that Mendel described were later given the name *genes*. Mendel recognized that a gene can exist in different forms.

Today these alternate forms are known as alleles. For example, pea seeds, the edible part of the plant we call peas, have a texture trait controlled by a single gene. This gene occurs in two alleles: one corresponding to round (smooth) peas, the other to wrinkled peas. Although an individual can carry only two alleles for a particular gene, each gene may have dozens of different alleles.

Mendel's experiments focused on interbreeding different strains of pea plants and then observing the traits that appeared in subsequent generations. When he crossbred plants with round peas and those with wrinkled peas, he discovered that all of the resulting offspring had round peas.

Today we know that peas are round and smooth when they contain the right amount of sugar. If peas are missing the gene that produces a protein called starch branching enzyme 1 (SBE1), the peas make too much sugar, causing the peas to swell and then wrinkle and shrivel as they dry. Mendel concluded that when an organism has two different alleles corresponding to the same genetic trait, one of the two may be dominant.

The other allele is said to be recessive, meaning that its presence will be detectable only if an organism has inherited the recessive gene from both parents. For convenience, geneticists designate alleles by a single letter—the dominant allele is represented by a capital letter and the recessive allele by a small letter. In the pea texture example, a plant inherits one allele for pea texture from each parent. The dominant allele that produces SBE1, resulting in round, smooth peas, is designated as *R*, while the recessive allele that does not produce SBE1 and produces wrinkled peas is designated as *r*.

To determine the set of alleles an organism has for a given trait just by visual observation can often be difficult. In the pea plant example, for instance, plants with smooth peas might be carrying two dominant alleles for that characteristic (*RR*) or one dominant and one recessive allele (*Rr*). Geneticists use the term *genotype* to refer to the combination of genes that code for a trait, while the term *phenotype* describes the physical manifestation of that trait. Therefore, the presence of two dominant alleles for pea texture (*RR*) would reflect the genotype while a smooth pea indicates the phenotype.

Mendel did not limit his experiments to testing the rules of inheritance of single traits. He also studied plant traits involving multiple pairs of genes, breeding plants that have round, yellow seeds with plants that produce wrinkled, green seeds. Such experiments demonstrated that the patterns of inheritance he observed in his experiments with single traits also apply to cases involving more complex gene combinations.

Exceptions to Mendel's Rules

Mendel published his studies in a science journal in 1865,

at which time no other scientist commented on his work. Since that time, geneticists have learned that sometimes genes do not easily conform to so-called Mendelian patterns of inheritance.

Incomplete Dominance

In cases of incomplete dominance, the inheritance of a dominant and a recessive allele results in a blending of traits to produce intermediate characteristics. For example, four-o'clock paint plants may have red, white, or pink flowers. Plants with red flowers have two copies of the dominant allele *R* for red flower colour (*RR*). Plants with white flowers have two copies of the recessive allele *r* for white flower colour (*rr*). Pink flowers result in plants with one copy of each allele (*Rr*), with each allele contributing to a blending of colors.

Quantitative Inheritance

Mendel focused his studies on traits determined by a single pair of genes, and the resulting phenotype was easy to distinguish. A tall plant can be markedly different from a short one, and a green pea can easily be distinguished from a yellow one. There are some traits, however, that are not easy to distinguish. Human skin colour, for example, may be any of a wide variety of shades. Traits such as skin colour differ from the ones Mendel studied because they are determined by more than one pair of genes. In this form of inheritance, known as quantitative inheritance, each pair of genes has only a slight effect on the trait, while the cumulative effect of all the genes determines the physical characteristics of the trait. At least four pairs of genes control human skin colour. Multiple genes also control many traits important in agriculture, such as milk production in cows and ear length in corn.

Multiple Alleles

Another exception to Mendelian genetics involves genes with multiple alleles. Certain traits are controlled by multiple alleles that have complex rules of dominance. In humans, for example, the gene for blood type has three alleles: I_A, I_B, and i.

With three alternatives for each member of a gene pair, there are six possible combinations of these genes (I_AI_A, I_BI_B, ii, I_Ai, I_Bi, I_AI_B). Although there are six possible combinations, humans have only four major blood types: A, B, AB, and O. This results because both I_A and I_B dominate over i, but not over each other, so a person with a gene combination of I_AI_A or I_Ai has blood type A. The gene combinations I_BI_B and I_Bi both produce blood type B. I_AI_B results in a blood type AB, and ii results in blood type O.

Gene Linkage

In his experiments, Mendel was careful to study traits in pea plants where one trait did not appear to influence another, such as the plant's height or the pea's texture. These two phenotypes (height and texture) occur randomly with respect to one another in a manner known as independent assortment. Today scientists understand that independent assortment occurs when the genes affecting the phenotypes are found on different chromosomes. An exception to independent assortment develops when genes appear near one another on the same chromosome. When genes occur on the same chromosome, they are inherited as a single unit.

Genes inherited in this way are said to be linked. For example, in fruit flies the genes affecting eye colour and wing length are inherited together because they appear on the same chromosome. But in many cases, genes on the same chromosome that are inherited together produce offspring with unexpected allele combinations. This results from a process called crossing over.

Sometimes at the beginning of meiosis, a chromosome pair (made up of a chromosome from the mother and a chromosome from the father) may intertwine and exchange sections of chromosome. The pair then breaks apart to form two chromosomes with a new combination of genes that differs from the combination supplied by the parents. Through this process of recombining genes, organisms can produce offspring with new combinations of maternal and paternal traits that may contribute to or enhance survival.

Sex-Linked Traits

Most chromosome pairs consist of identical, or homologous, partners. In many species, including humans, there is one pair of chromosomes in which the partners noticeably differ from each other. These are called the sex chromosomes because they determine the differences between males and females. Genes located on the sex chromosomes display different patterns of inheritance than genes located on other chromosomes.

In human females, the sex chromosomes consist of two X chromosomes, while males have an X chromosome and a shorter Y chromosome with many fewer genes. In males the X chromosome contains many genes that have no corresponding gene on the Y chromosome. A male's X chromosome may contain a recessive allele associated with a genetic disorder, such as hemophilia or Duchenne muscular dystrophy.

In this case, males do not have a normal second copy of the gene on the Y chromosome to mask the effects of the recessive gene, and disease typically results. Additional examples of sex-linked traits include red-green colour blindness in humans and eye colour in fruit flies.

THE GENETIC CODE

The structure of DNA encodes all the information every cell needs to function and thrive. In addition, DNA carries hereditary information in a form that can be copied and passed intact from generation to generation. A gene is a segment of DNA.

The biochemical instructions found within most genes, known as the genetic code, specify the chemical structure of a particular protein. Proteins are composed of long chains of amino acids, and the specific sequence of these amino acids dictates the function of each protein.

The DNA structure of a gene determines the arrangement of amino acids in a protein, ultimately determining the type and function of the protein manufactured.

DNA Structure

DNA molecules form from chains of building blocks called nucleotides. Each nucleotide consists of a sugar molecule called deoxyribose that bonds to a phosphate molecule and to a nitrogen-containing compound, known as a base. DNA uses four bases in its structure: adenine (A), cytosine (C), guanine (G), and thymine (T). The order of the bases in a DNA molecule—the genetic code—determines the amino acid sequence of a protein.

In the cells of most organisms, two long strands of DNA join in a single molecule that resembles a spiraling ladder, commonly called a double helix. Alternating phosphate and sugar molecules form each side of this ladder. Bases from one DNA strand join with bases from another strand to form the rungs of the ladder, holding the double helix together.

The pairing of bases in the DNA double helix is highly specific—adenine always joins with thymine, and guanine always links to cytosine. These base combinations, known as complementary base pairing, play a fundamental role in DNA's function by aiding in the replication and storage of genetic information.

Complementary base pairing also enables scientists to predict the sequence of bases on one strand of a DNA molecule if they know the order on the corresponding, or complementary, DNA strand. Scientists use complementary base pairing to help identify the genes on a particular chromosome and to develop methods used in genetic engineering.

Genes line up in a row along the length of a DNA molecule. In humans a single gene can vary in length from 100 to over 1,000,000 bases. Genes make up less than 2 percent of the length of a DNA molecule. The rest of the DNA molecule is made up of long, highly repetitive nucleotide sequences. Once dismissed as "junk" DNA, scientists now believe these nucleotide sequences may play a role in the survival of cells. Identifying the function of these sequences is a thriving field of genetics research.

DNA Replication

In order for inherited traits to be transmitted from parent to child, the genetic information encoded in DNA must be copied with great precision during cell division. The accuracy of DNA replication depends upon the complementary pairing of bases. During replication, the DNA double helix unwinds and bonds joining the base pairs break, separating the DNA molecule into two separate strands. Each strand of DNA directs the synthesis of another complementary strand. The unpaired bases of each DNA strand attach to bases floating within the cell. But the DNA strand's unpaired bases bond only with specific, complementary bases—for example, an adenine base will bond only with a thymine base and a cytosine bases will pair only with a guanine base.

Once all of the bases of a DNA strand bond to complementary bases, the complementary bases then link to each other, forming a new DNA double-helix molecule. Thus the original DNA molecule replicates into two DNA molecules that are exact duplicates.

PROTEIN SYNTHESIS

DNA replication ensures that the genetic instructions encoded in DNA can be used continuously through generations to produce the proteins that build and operate the cells of an organism. The process of tapping the genetic code to create proteins, known as protein synthesis, has two crucial steps: transcription and translation.

Transcription

Transcription transfers the genetic code from a molecule of DNA to an intermediary molecule called ribonucleic acid (RNA). The basic nucleotide structure of RNA resembles that of DNA, but the two compounds have three critical differences. First, the structure of RNA incorporates the sugar ribose rather than deoxyribose, the sugar in DNA. Second, RNA uses the base uracil (U) instead of thymine (T). In RNA uracil binds with adenine just as thymine does in DNA. Third, RNA usually exists as a single strand, unlike the double-helix structure that normally characterizes DNA.

Transcription involves the production of a special kind of RNA known as messenger RNA (mRNA). The process begins when the two strands of a DNA molecule separate, a task directed by the enzyme RNA polymerase. After the double helix splits apart, one of the strands serves as a template, or pattern, for the formation of a complementary mRNA molecule. Free-floating individual bases within the cell bind to the bases on the DNA template using complementary base pairing. The individual bases then link together to form a strand of mRNA.

In eukaryotes (organisms whose cells have a nucleus), the mRNA strand undergoes an additional step before the next stage of protein synthesis can occur. The mRNA strand consists of coding regions called exons separated by regions called introns. The introns do not contribute to protein synthesis. Special enzymes in the nucleus remove the introns from the mRNA strand. The remaining exons then link together to form an mRNA strand that contains the entire code for making a protein.

Translation

Once transcription is complete and the genetic code has been copied onto mRNA, the genetic code must be converted into the language of proteins. That is, the information coded in the four bases found in mRNA must be translated into the instructions encoded by the 20 amino acids used in the formation of proteins. This process, called translation, takes place in cellular organelles called ribosomes. In eukaryotes, mRNA travels out of the nucleus into the cell body to attach to a ribosome.

In prokaryotes (organisms without a nucleus), the ribosome clasps mRNA and starts translation before these strands have finished transcription and separated from the DNA. In both eukaryotes and prokaryotes, the ribosome acts like a workbench and clamp that holds the mRNA strand and coordinates the activity of enzymes and other molecules essential to translation. Another form of RNA called transfer RNA (tRNA) is found in the cytoplasm of the cell.

There are many different types of tRNA, and each type binds with one of the 20 amino acids used in protein formation. One end of a tRNA binds with a specific amino acid. The other end carries three bases, known as an anticodon. The tRNA with an amino acid attached travels to the ribosome where the mRNA is stationed. The anticodon of the tRNA undergoes complementary base pairing with a series of three bases on the mRNA, known as the codon. The mRNA codon codes for the type of amino acid carried by the tRNA.

A second tRNA bonds with the next codon on the mRNA. The resident tRNA transfers its amino acid to the amino acid of the incoming tRNA and then leaves the ribosome. This process continues repeatedly, with new tRNA receiving the growing chain of amino acids, known as a polypeptide chain, from a resident tRNA. The ribosome moves the mRNA strand one codon at a time, making new codons available to bind with tRNAs. The process ends when the entire sequence of mRNA has been translated. The polypeptide chain falls away from the ribosome as a newly formed protein, ready to go to work in the cell.

MUTATIONS

Occasionally mistakes occur during DNA replication and protein synthesis. Any alteration in the structure of a gene results in a mutation. Mutations occur during DNA replication when the chemical structure of genes undergoes random modifications. Once a change has occurred, the altered genes continue to replicate in their changed form unless another mutation occurs. Sometimes mutations occur during transcription or translation, causing protein synthesis to go awry. Although mutations may occur in any living cell, they are most important when they occur in gametes because then the change affects the traits of following generations.

Most mutations harm an organism. If a mutation occurs in a gene sequence that codes for a particular protein, the mutation may result in a change in the amino acid sequence directed by the gene. This change, in turn, may affect the function of the protein. The implications can be significant: The amino acid sequence distinguishing normal hemoglobin from the altered form of hemoglobin responsible for sickle-cell anemia differs by only a single amino acid.

Some mutations may be neutral or silent and do not affect the function of a protein. Occasionally a mutation benefits an organism. Over the course of evolutionary time, however, mutations serve the crucial role of providing organisms with previously nonexistent proteins. In this way, mutations are a driving force behind genetic diversity and the rise of new or more competitive species better able to adapt to changes, such as climate variations, depletion of food sources, or the emergence of new types of disease.

Mutations can produce a change in any region of a DNA molecule. In a *point mutation,* for example, a single nucleotide replaces another nucleotide. Although a point mutation produces a small change to the DNA sequence, it may cause a change in the amino acid sequence, and thus the function, of a protein.

Far more serious are mutations that involve the addition or deletion of one or more bases from a DNA molecule. Adding or subtracting even a single base from a normal sequence during transcription can disrupt translation by shifting the "reading frame" of every subsequent codon. For example, an mRNA strand may include two codons in the following sequence: AUG UGA. The addition of a cytosine base at the beginning of this sequence shifts the "spelling" of these codons so that they read: CAU GUG. This may result in an incorrect amino acid sequence during translation, or the protein may be truncated. Known as *frameshift mutation,* this type of alteration could result in the production of a protein with no real function or one with a harmful effect.

Sometimes mutations are caused by *transposition,* in which long stretches of DNA (containing one or more genes) move from one chromosome to another. These jumping genes, called transposons, can disrupt transcription and change the type of amino acids inserted into a protein. Transposons rearrange and interrupt genes in a way that generally improves the genetic variation of a species.

While mutations can occur spontaneously, some can be caused by exposure to physical or chemical agents in the environment called mutagens. Common environmental mutagens include ultraviolet rays from the sun and various chemicals, such as asbestos, cigarette smoke, and nitrous acid. High-energy radiation, such as medical X rays, can cause DNA strands to break, leading to the deletion of potentially important genetic information.

Radiation damage can also affect an entire chromosome, disrupting the function of many genes. In chromosomal translocation, a piece of one chromosome breaks off and merges with another chromosome. In some cases, large sections of chromosomes may break off and be lost.

The cell has highly effective self-repair mechanisms that can correct the harmful changes made by mutations and prevent some mutations from being passed on. Some 50 specialized enzymes locate different types of faulty sequences in the DNA and clip out those flaws. Another repair mechanism scans DNA after replication and marks mismatched base pairs for repair.

GENE REGULATION

The processes that enable information to be copied from genes and then used to synthesize proteins must be regulated if an organism is to survive. Different cells within an organism share the same set of chromosomes. In each cell some genes are active while others are not.

For example, in humans only red blood cells manufacture the protein hemoglobin and only pancreas cells make the digestive enzyme known as trypsin, even though both types of cells contain the genes to produce both hemoglobin and trypsin. Each cell produces different proteins according to its needs so that it does not waste energy by producing proteins that will not be used.

A variety of mechanisms regulate gene activity in cells. One method involves turning on or off gene transcription, sometimes by blocking the action of RNA polymerase, an enzyme that initiates transcription. Gene regulation may also involve mechanisms that slow or speed the rate of transcription, using specialized regulatory proteins that bind to DNA. Depending on an organism's particular needs, one regulatory protein may spur transcription for a particular protein, and later, another regulatory protein may slow or halt transcription.

Prokaryotes

The bacterium *Escherichia coli* (commonly referred to as *E. coli*), found in the intestines of humans and other mammals, provides a good example of gene regulation. *E. coli* uses three enzymes to digest lactose, the primary sugar found in milk. The bacterium produces great quantities of lactose-digesting enzymes when lactose is present and saves energy by not synthesizing the enzymes when the sugar is not available. *E. coli* prefers consuming glucose to lactose, so the bacterium produces these enzymes in the presence of lactose only when no glucose is available.

A region of DNA known as an operon controls this gene regulation process. In *E. coli*, the operon includes at least five genes: Three genes, called lac genes, code for the enzymes that digest lactose; one gene encodes for a regulatory protein, called a repressor, that can sense the presence or absence of lactose; and one gene, called an operator, activates transcription, the first step in the synthesis of the lactose-digesting enzymes.

In the absence of lactose, the repressor protein binds with the operator gene to block transcription. This prevents the lac genes from being transcribed and halts production of the lactose-digesting enzymes. If lactose is present, the repressor protein binds to the sugar, leaving the operator gene free to trigger transcription of the lac genes. The transcription of the lac genes produces the mRNA that will direct the production of the three lactose-digesting enzymes.

Eukaryotes

Gene regulation in eukaryotes is more complex than in

bacteria and other prokaryotes. Even the simplest eukaryotes have far more genes than prokaryotes, and these genes must be turned on or off as conditions dictate. Most multicellular organisms contain different types of cells that serve specialized functions. The cells of an animal's heart, blood, skin, liver, and muscles all contain the same genes. But in order to carry out their specific functions within the body, each cell must produce different proteins and respond to changing environmental stimuli, such as glucose levels in the blood or body temperature. Such specialization is possible only with sophisticated gene regulation.

Eukaryotes use a variety of mechanisms to ensure that each cell uses the exact proteins it needs at any given moment. In one method, eukaryotic cells use DNA sequences called enhancers to stimulate the transcription of genes located far away from the point on the chromosome where transcription occurs. If a specific protein binds to an enhancer site on the DNA, it causes the DNA to fold so that the enhancer site is brought closer to the site where transcription occurs. This action can activate or speed up transcription in the genes surrounding the enhancer site, thereby affecting the type and quantity of proteins the cell will produce. Enhancers often exert their effects on large groups of related genes, such as the genes that produce the set of proteins that form a muscle cell.

Gene regulation can also take place after transcription has occurred by interfering with the steps that modify mRNA before it leaves the nucleus to take part in translation. This process typically involves removing exons (segments that code for specific proteins) and introns. These sections of the mRNA can be modified in more than one way, enabling a cell to synthesize different proteins depending on its needs.

GENES IN DEVELOPMENT

Gene regulation helps individual cells within an organism function in a specialized way. Other regulatory mechanisms coordinate the genes that determine how cells develop. All of the specialized cells in an organism, including those of the skin, muscle, bone, liver, and brain, derive from identical copies of a single fertilized egg cell. Each of these cells has the exact same DNA as the original cell, even though they have vastly different appearances and functions. Genes dictate how these cells specialize.

Early in an organism's embryonic development the overall body plan forms. Individual cells commit to a particular layer and region of the embryo, often migrating from one location to another to do so. As the organism grows, cells become part of a particular body organ or tissue, such as skin or muscle. Ultimately, most cells become highly specialized—not only to develop into a neuron rather than a muscle cell, for example, but to become a sensory neuron instead of a motor neuron. This process of specialization is called differentiation. At each stage of the differentiation process, specific genes known as developmental control genes actively turn on and switch off the genes that differentiate cells.

One class of developmental control genes, known as homeotic genes, directs the formation of particular body parts. Activating one set of homeotic genes instructs part of an embryo to develop into a leg, for example, while another set initiates the formation of the head. If a homeotic gene becomes altered or damaged, an organism's body development can be dramatically disrupted. A change in a single gene in some insects, for instance, can cause a leg to grow where an antenna belongs.

Homeotic genes work by regulating the activity of other genes. Homeotic genes code for the production of a regulatory protein that can bind to DNA and thus affect the transcription of one or more genes. This enables homeotic genes to initiate or halt the development and specialization of characteristics in an organism.

Nearly identical homeotic genes have been identified in varied organisms, such as insects, worms, mice, birds, and humans, where they serve similar embryonic development functions. Scientists theorize that homeotic genes first appeared in a single ancestor common to all these organisms. Sometime in evolutionary history, these organisms diverged from their common ancestor, but the homeotic genes continued to be passed down through generations virtually unchanged during the evolution of these new organisms.

SCIENTISTS WORK WITH GENES

Scientists have developed a number of biochemical and genetic techniques by which DNA can be separated, rearranged, and transferred from one cell to another. Some of these laboratory methods help scientists study the properties of genes in nature—for example, by comparing DNA from different animals to find out whether those animals are closely related to each other or only distant relatives. Other DNA techniques provide tools for genetic engineering—the alteration of genes in an organism. These tools are used in industry to develop commercial products, such as hardier crops, microbes that can break down oil slicks or decompose garbage, and improved medicines.

RECOMBINANT DNA

The DNA molecules of all life forms, from oak trees to sea horses, have the same structure and the same four bases. Scientists have made use of these similarities in a technology called recombinant DNA. In this laboratory method, one or more genes of an organism are introduced into a second organism. The new genes, sometimes known as foreign DNA, become functional in the second organism and produce a desired protein. In this way, scientists can create changes in the genetic makeup of an organism that would be unlikely to occur through natural processes.

Scientists use recombinant DNA when they want to obtain large amounts of a protein, such as insulin, produced by a gene. Insulin was once in short supply for diabetics, whose bodies lack adequate supplies. Insulin supplies were derived from cows in an expensive and time-consuming process. Today recombinant DNA techniques produce insulin cheaply and in abundance. The first step in creating insulin using recombinant DNA is to isolate the sequence of nucleotides in the DNA of a human cell that forms the insulin gene. Scientists use *restriction enzymes,* specialized proteins that act like molecular scissors, to cut the double-stranded DNA at the point where the insulin gene occurs. The isolated DNA can then be recombined, or spliced, with a vector, a fragment of DNA that is able to transport genes from one organism to another. A vector may be a plasmid, a small, circular segment of DNA found in bacteria.

Bacteriophages, viruses that are parasites of bacteria, also act as vectors. Scientists insert the vector containing the insulin gene into a bacterium, such as *E. coli*. Within just a few hours, a single *E. coli* will reproduce hundreds of times to make millions of cells, all containing exact copies of the insulin-producing gene inserted by the scientists. This process of making many cells with identical DNA is known as cloning.

DNA LIBRARIES

A DNA library is a storehouse of genetic information maintained in bacteria instead of books. These bacteria are clones created by recombinant DNA, and the foreign DNA they hold is the library's store of information. DNA libraries are helpful to scientists who require a plentiful supply of particular DNA segments to do their work. These repositories of genetic information are stored in small tubes, which can easily be shipped to other researchers for study.

Each library has a unifying theme. For example, a library may contain the entire chromosomal DNA, or genome, of a given organism, or it may consist of genes that are active within certain types of cells, such as heart cells. To create a library of the human genome, DNA from all the human chromosomes would be cut into many pieces. These pieces would be randomly inserted into vectors, such as plasmids, which would then be placed into a population of bacteria. Taken together, the entire population of bacteria would contain all the DNA of the human chromosomes.

POLYMERASE CHAIN REACTION

Polymerase chain reaction (PCR) offers an alternative to vector-based cloning as a means of generating numerous copies of DNA from a small initial sample. Performed in a test tube, PCR mirrors the way in which DNA is replicated within a cell. To perform PCR, scientists isolate the piece of DNA to be amplified (multiplied) in a test tube and heat it to separate the two strands of the molecule. As cooling occurs, short pieces of DNA called primers are added to the test tube.

The primers attach to each strand, marking the segment that will be cloned. Free-floating nucleotides and an enzyme called DNA polymerase are then added to the mixture. DNA polymerase uses the free-floating nucleotides to build a complementary copy of each amplified DNA segment, resulting in two new double-stranded DNA molecules. Each cycle of heating and cooling doubles the amount of the desired DNA fragment in the test tube. In a matter of hours, scientists can obtain millions of copies of a desired piece of DNA. PCR enables scientists to amplify traces of DNA found at a crime scene or in a fossil animal to produce sufficient quantities to study.

GEL ELECTROPHORESIS

PCR and recombinant DNA techniques create large amounts of DNA segments. To study the structure of these segments, researchers use a process known as gel electrophoresis. This technique can be used to identify genes in humans that have previously been identified in other organisms, such as fruit flies. It can also be used to compare the DNA found from blood or hair samples at a crime scene with the DNA of a suspect in the crime. In gel electrophoresis, restriction enzymes break up the DNA under study into restriction fragments of varying lengths. Solutions containing these fragments are placed within a thick gel.

An electric current is applied to the gel, causing one end of the gel to have a positive charge and the other to have a negative charge. All of the restriction fragments begin to move from the negative end of the gel toward the positive end. The smaller fragments move faster than the larger fragments. When the current shuts off, typically after several hours, the DNA fragments have spread out across the gel, with the smaller ones closer to the positive end. The dispersed fragments display a pattern resembling a bar code.

Each bar in this pattern contains DNA fragments of a certain size. Scientists can identify specific restriction fragments by their location on the gel. A complementary sequence of DNA can be used as a probe to find a restriction fragment on the gel that has a particular nucleotide sequence. Scientists may use DNA found in blood at a crime scene as the probe to see if it pairs up with any of the DNA fragments in the gel electrophoresis. If pairing occurs, the DNA from the crime scene is from the same person who provided the DNA sample for the gel electrophoresis.

DNA SEQUENCING

Once an interesting piece of DNA has been isolated or identified, scientists often need to determine if the sequence of nucleotides in the fragment is related to known genes and to determine what kind of protein it might make. Scientists use DNA sequencing to detect genetic mutations linked to diseases such as cystic fibrosis. Scientists have also used this method to alter the sequence of a gene and study the function of the resulting protein. In DNA sequencing, scientists create many copies of a single-stranded DNA fragment that will be used to synthesize a new DNA strand.

An equal number of copies of the fragment are placed into four different test tubes to act as the template for the synthesis of a new strand. The enzyme DNA polymerase and free nucleotides are added to each test tube. Each test tube also receives one type of dideoxy nucleotide—a nucleotide that closely resembles either adenine, guanine, thymine, or cytosine. These nucleotides can attach to the end of the new complementary DNA strand, but they cannot bind to anything else, thus they terminate the synthesis of the new DNA strand.

DNA polymerase uses the free nucleotides to build a complementary DNA strand. If the original DNA fragment contains guanine, DNA polymerase delivers a cytosine dideoxy nucleotide to pair with the guanine base on the original strand. The cytosine links with the growing chain of nucleotides on the complementary DNA strand, but it is unable to bind with any other nucleotide. The newly formed DNA fragment terminates with the cytosine dideoxy nucleotide at the end of the chain. The reactions in each of the four test tubes produce a series of DNA fragments in which the new strands terminate at a known base.

Each test tube produces fragments that differ in length from the other test tubes. The newly formed fragments are sorted in an electrophoresis gel that can detect differences as small as one nucleotide in length. By analyzing these sorted fragments, scientists can determine the complementary base sequence for the original DNA fragment. This sequencing method has become a routine laboratory technique, automated with specialized machines and computers that can prepare DNA samples and read nucleotide sequences far faster and more accurately than people can.

GENE CHIP

The gene chip, also known as a DNA chip or DNA microarray, is a thumbnail-sized chip of glass or silicon that carries DNA instead of electronic circuits. Gene chips can identify the genes that are active within a cell and help identify mutated genes. In one application, scientists take a single strand of DNA that contains a defective gene and use ultraviolet light to attach the strand onto a glass or silicon chip.

A second DNA strand isolated from a patient is attached to fluorescent markers and deposited onto the chip. If the patient's DNA strand bonds with the DNA already bonded to the chip, then the individual's DNA contains the defective gene.

When the DNA on the chip pairs with the fluorescent DNA, it develops a fluorescent glow that can be viewed with a microscope and interpreted by a computer. A diagnostic gene chip may soon be manufactured to hold the DNA sequences of all the known disease-causing genes, making diagnosis for genetic disorders fast, reliable, and inexpensive. Gene chips also distinguish between active DNA—DNA that is being transcribed to produce mRNA—and inactive DNA.

Researchers use these chips to learn how the transcription of a group of genes is affected when cells are exposed to a drug.

HUMAN GENETICS

Our understanding of human genetics builds on a foundation of information obtained from studying other organisms. Until the 1980s, genetic researchers focused their work on the fundamental genetic processes in simpler organisms, such as bacteria, plants, and fruit flies. Today an expanded array of tools available for the direct study of human genetics attracts scientists from around the world to collaborate to identify and study every human gene.

The genetic principles that Mendel first discovered in plants apply to humans as well. As in all other life forms, the DNA found in human cells encodes the proteins that are essential for reproduction, survival, and growth. The unique structure and behaviour of DNA ensures that human traits are passed from generation to generation and accounts for why parents, children, and grandchildren often have similar facial features, hair colour, height, and athletic or artistic abilities. Yet each of us inherits a unique genetic legacy from our parents and more distant ancestors. With the exception of identical twins, no two people have the exact same combination of alleles for the estimated 20,000 to 25,000 human genes.

Some human traits are controlled largely by a single gene. But most inheritable characteristics are influenced by a number of genes that interact in a complex fashion. Also, personal experiences and environmental factors combine with genetic influences to shape certain traits, including vulnerability to disease and characteristics such as intelligence, emotions, talents, and personality.

Human Genome

Human genes reside on 23 pairs of chromosomes found in the nucleus of every body cell except gamete cells. In each pair, one of the chromosomes is inherited from the mother and the other is passed down from the father. About 2 m(7 ft) of DNA is packaged into each chromosome. All of the genes carried on chromosomes form the human genome. A lesser amount of DNA can be found in mitochondria, cellular organelles responsible for creating the energy used in cell activities.

All but one of these 23 pairs are composed of chromosomes nearly identical in shape. Each of these 22 chromosome pairs, known as autosomes, contains the same genes (although they likely carry different alleles). The autosome pairs vary considerably in length, and scientists number them according to their relative size: Pair number 1 is the longest pair and pair number 22 is the shortest.

Rounding out the human genome is the 23rd pair of chromosomes, known as the sex chromosomes, which determine the sex of an individual. Females inherit two X chromosomes, a matched pair carrying the same genes. One X chromosome is inherited from the mother and one X chromosome is inherited from the father. Males inherit an X chromosome from their mother and a Y chromosome from their father. The Y chromosome is shorter than the X chromosome and bears far fewer genes. One gene on the Y chromosome causes an embryo to develop as a male rather than a female.

Humans produce gamete cells for sexual reproduction. These gametes contain a haploid number of chromosomes—23 chromosomes instead of the full complement of 46. Female gamete cells mature into eggs, with each egg containing chromosomes 1 through 22 and an X chromosome. Males produce gametes that mature into sperm, and each sperm cell has a single set of chromosomes 1 through 22 and either an X or a Y chromosome. During fertilization, an egg that joins with a sperm containing a Y chromosome develops into a male, and an egg fertilized by a sperm containing an X chromosome develops into a female.

Human Genetic Disorders

Thousands of inherited diseases caused by altered genes and chromosomal abnormalities affect humans. These disorders cause problems such as physical deformities, metabolic dysfunction, and developmental problems. Medical surveys indicate that roughly 1 percent of newborns in the United States have a single-gene defect. As many as 1 baby in 200 is born with a chromosomal abnormality serious enough to produce physical defects or mental retardation.

It is misleading to say that a person "inherits the gene" for a disease, since humans are born with the same number and types of genes. We inherit allele forms of specific genes, and these alleles may be defective. Most of the known inherited genetic disorders are caused by the mutation of a single gene, resulting in alleles that produce disease. These defects often produce disturbances in the body's biochemical processes, such as inhibiting the action of an important enzyme or stimulating the overproduction of a harmful substance. Frequently the consequences of such problems can cause severe disability or be fatal.

Many single-gene disorders follow Mendelian patterns of inheritance. A mother and father each pass an allele for a specific gene on to a child. If one of the alleles is defective and causes disease, the child will develop the disease according to a dominant-recessive pattern of inheritance. For example, cystic fibrosis (CF), a metabolic disorder that causes a progressive loss of lung function, is caused by a mutation in the recessive allele of a gene responsible for regulating salt content in the lungs.

The recessive allele is unable to direct the production of a key protein, resulting in a salt imbalance that causes thick, suffocating mucus to build up in the lungs. If a baby inherits the defective allele from just one parent, no disease results. But the infant who inherits the defective allele from both parents will be born with the disease.

In other cases, a single dominant allele causes genetic disease. Huntington's disease, a condition characterized by involuntary movements, dementia, and eventually death, is caused by the inheritance of a pair of alleles in which a defective allele dominates the normal allele for the gene. An affected parent has a 50 percent chance of passing the defective allele to a child. A child who inherits the dominant defective allele from just one parent will develop the disease.

Other inherited genetic diseases are caused by defects in the genes found on the X chromosome. Hemophilia, the inability of the blood to clot and heal a wound, is caused by a defect in an allele located on the X chromosome that helps produce proteins involved in the clotting process. Women who inherit this defective allele usually have the normal allele on their second X chromosome, which produces enough of these clotting proteins for the body to remain healthy.

Women who inherit this faulty allele have a 50 percent chance of passing the defective allele on to their children. Males who inherit this defective allele do not have a normal version of the allele on their Y chromosome and so cannot produce clotting proteins to heal wounds. Hemophiliacs are almost always males who have inherited an X chromosome with the faulty allele from their mother.

Other genetic disorders arise due to the inheritance of an abnormal number of chromosomes or a defective chromosome structure. These chromosomal abnormalities have a devastating impact: Many fetuses with such defects, particularly those with missing chromosomes, will die prenatally, resulting in miscarriage (spontaneous abortion). In other cases, newborns with chromosomal abnormalities suffer from physical problems or varying degrees of mental retardation. Down syndrome occurs when an individual's cells carry an extra copy of chromosome 21. People born with this condition have characteristic facial features, short stature, severe developmental disabilities, and a shortened life expectancy.

GENETICS AND CANCER

Cancer is a common name for many diseases that affect different body tissues, including the skin and the liver. All cancers involve alterations in genes that control cell division. These alterations cause cells to replicate abnormally and form tumors. Cancers generally arise from mutations that occur directly in the somatic cells, any cells of the body with the exception of the gametes (sperm and egg cells). Since the genetic mutations have not occurred in gametes, the mutations are not inherited by the next generation.

While cancer is not a traditional inherited genetic disorder, scientists have determined that a genetic component plays a strong role in the development of the disease. Geneticists have identified many different genes with certain alleles that appear to increase an individual's susceptibility to cancer. A notable example involves two genes linked to breast cancer. Researchers estimate that more than half of the women with a family history of breast cancer who inherit mutated alleles of these two genes, known as BRCA1 and BRCA2, will develop breast cancer by the age of 70. In contrast, women who lack either of the mutated alleles have only a 13 percent chance of developing the disease.

For many cancers, researchers believe that mutations in several different genes must accumulate before cancer develops. As a person ages, errors in DNA replication may occur during cell division, or cells may be damaged by exposure to certain environmental factors, including cigarette smoke, radiation, and chemical pollutants. As a result, an accumulation of mutations may develop in two types of genes: tumor suppressor genes and oncogenes. Tumor suppressor genes normally function to halt cell division, while oncogenes function to activate cell division. A mutation in either type of gene can stimulate nonstop cell division. These types of defects have been linked to some cases of leukemia as well as to cancers of the ovaries, lungs, colon, and other organs.

Genetics and Aging

A growing area of study focuses on the link between aging and genetics. Scientists have determined that structures called telomeres, long, repetitive sequences of nucleotides at the end of chromosomes, affect the aging process. Each time a cell divides, telomeres become shorter. When the structures shorten to a certain length, the process of cell division terminates. The cells of these chromosomes continue to live, but they never divide again.

Laboratory tests suggest that an enzyme produced in gamete cells of the human body, called telomerase, can maintain telomere length in human cells, enabling them to continue dividing, perhaps indefinitely.

Scientists hoping to lasso the life-extending properties of telomerase have been confounded by research indicating that telomerase is also active in rapidly dividing cancer cells. Before telomerase can be used to slow or halt aging, scientists must learn how to manipulate the enzyme so that it does not promote cancer growth.

Genes and Behaviour

Scientists actively explore the links between genes and behaviour to determine both the patterns and the limits of genetic influence. Such studies continue to be controversial because behaviour or mental processes can be difficult to measure objectively. Furthermore, many behavioral traits, both normal and abnormal, are complex, influenced by many genes as well as by personal experiences.

Studies of the possible genetic components of psychiatric disorders have yielded mixed results. Geneticists have identified at least two genes linked to schizophrenia, a condition characterized by hallucinations, delusion, paranoia, and other symptoms. Other studies that reported the discovery of genes that influence bipolar disorder (also known as manic-depressive illness) and alcoholism have been reversed or questioned. Though attempts to identify genes linked to these disorders have been flawed, scientists have little doubt that the conditions do have a genetic component.

Scientists have established links between genes and certain antisocial or violent behaviors. For instance, researchers have identified a gene on the X chromosome that has been tied to extremely violent behaviour in men.

They identified the gene in members of several families with a multigenerational history of violent, criminal behaviour. They identified gene codes for monoamine oxidase inhibitor (MAO), an enzyme that helps nerve cells in the brain communicate with each other. Males in an affected family who inherit a defective allele for MAO do not produce enough of the enzyme. As a result, low levels of MAO change the activity of certain brain nerve cells, possibly contributing to socially unacceptable behaviour.

While research suggests that men who have a defective allele for the MAO gene are more prone to aggressive behaviour, experts cite numerous reasons for concern and doubt. Few scientists believe that a single gene could have a leading role in influencing complex behaviors. Others charge that these kinds of investigations promote an unreasonably simplistic view of genetic determinism, in which genes can be blamed for certain behaviors.

Critics note that studies have identified many men who carried the defective allele for MAO production and never committed a violent act. Clearly, nongenetic influences—as varied as an individual's family life, work circumstances, attitudes, diet, and emotional state—affect complex behaviors.

IDENTIFYING GENETIC DISORDERS

Health-care professionals who specialize in genetic disorders use a variety of methods to identify inherited conditions. Analysis of a family medical history, known as pedigree analysis, is used to track the transmission of a condition through generations. Blood tests that identify specific DNA sequences can reveal carriers of a disease-causing gene who have no symptoms of the disease.

Geneticists collect a person's medical family history to trace the inheritance of a genetic trait among multiple generations. The information is placed in a pedigree, which resembles a traditional multigenerational family tree but includes information about individuals who were diagnosed with a particular disorder or who suffered from certain medical symptoms. A pedigree can help researchers recognize diseases that express themselves in dominant or recessive alleles.

Dominant disorders affect every generation. Recessive disorders may cluster in a single generation, reflecting when two parents who both carry a recessive allele for a disease have one or more children who develop the disease. A pedigree can also identify diseases that show X-linked inheritance. Pedigree analysis can be useful when combined with certain genetic tests. A blood sample taken from a person who is at risk for a genetic disorder can be compared with a DNA sequence known to cause the disorder in question.

Other genetic tests can reveal if a person has extra chromosomes, missing chromosomes, or chromosomes that have attached to one another in unusual ways. In some cases, these chromosomal abnormalities may produce genetic disorders in children or they may affect a person's ability to conceive a child. Genetic testing can also identify disorders in a fetus, enabling parents to learn early in a pregnancy if a fetus will likely be born with health problems or develop them later in life.

Presymptomatic testing can identify DNA abnormalities in a person before health problems develop. In the case of certain inherited heart conditions, for example, these tests enable a person to make healthy lifestyle changes or take other preventative measures, such as medications, to lower the risk of illness or death.

Medical genetic testing raises challenging issues because such tests typically provide statistical possibilities rather than a definite prediction of whether a person will develop a given genetic disease. A test result may indicate, for example, that a person has a 75 percent risk of developing colon cancer by the age of 65. Such results enable a physician to perform appropriate screening tests on the at-risk person in order to identify the disease at its earliest stages, when it is most treatable. At the same time, however, physicians must decide at what age the person's screening should begin and whether the benefits of early screening are worth the drawbacks of frequent screening.

These drawbacks include expense, patient anxiety and discomfort, and exposure to radioactivity or other harmful substances used in testing. Different problems are posed by genetic screening tests that diagnose conditions for which no preventive measures exist, such as Alzheimer's disease, a progressive brain disorder that causes the loss of mental function. People may find it devastating to learn that they are at risk for a deadly disease that cannot be prevented by medical measures or lifestyle choices.

GENE THERAPY

A recent development in genetic technology known as gene therapy focuses on curing inherited disorders. In experiments using gene therapy, researchers have replaced defective genes with normal alleles, inactivated a mutated gene, or inserted a normal form of a gene into a chromosome. The earliest success in human gene therapy involved the treatment of infants who cannot produce adenosine deaminase (ADA), an enzyme important to normal function of the immune system. Scientists have successfully inserted the normal allele for the gene that codes for the enzyme into cells in ADA-deficient children. Preliminary evidence indicates that this gene therapy leads to better immune function in recipients. Researchers are also exploring gene therapy's potential to help treat people with many other conditions, including certain cancers, hemophilia, heart disease, and cystic fibrosis.

Although the United States Food and Drug Administration (FDA) has approved more than 400 clinical trials in gene therapy, this method of treating disease remains far from an unqualified medical success. Treatments usually produce some improvement in the underlying condition, but not enough to consider the therapy suitable for large-scale use. The death of a patient involved in a gene therapy experiment in 1999 caused the National Institutes of Health (NIH), a federal agency that monitors gene therapy studies, to reevaluate the safety and effectiveness of gene therapy clinical trials.

HUMAN GENOME PROJECT

The Human Genome Project is the most ambitious project in the history of biology. The program's challenging goal was to identify and sequence all of the DNA in human chromosomes. The project was initiated in 1990 in the United States with government funding, and it rapidly grew into an international consortium of academic centers and drug companies in China, France, Germany, Japan, the United Kingdom, and the United States. The consortium initially hoped to reach its goal by the year 2005.

In 1998 Celera Genomics, a privately funded biotechnology firm, announced that it would sequence the human genome by the year 2000 using different sequencing strategies than those used by the public consortium. This announcement triggered a heated race between Celera Genomics and the public consortium to complete the genome project. In June 2000 both teams declared victory when they jointly announced that they had separately completed a rough draft of the genome. The two teams published their findings simultaneously, although in two different journals, in February 2001. The draft provided a basic outline of 90 percent of the human genome. Scientists from the public consortium completed the final sequencing of the human genome in April 2003, two years earlier than planned.

The completed human genome has provided scientists with a detailed blueprint of our complex genetic code. Large computer databases of genetic information enable scientists to look for patterns and relationships among the actions of different genes. Among the findings about the human genome was that the number of genes in the human genome is much lower than was predicted—only about 20,000 to 25,000 genes compared to the expected 100,000 genes. This number is a little more than twice the number of genes found in the fruit fly.

Scientists are now turning their attention to studying how the relatively low number of genes in the human genome can produce the complex structures found in humans. Scientists have long known that a single gene produces a single protein and that this single protein subsequently may be processed into several different proteins. In a new science known as proteomics, scientists seek to identify and understand the function of all the proteins in the human body. They theorize that there may be many more proteins than there are genes—that is, more than 25,000.

Among other advances, the database of proteins derived from proteomics is expected to help scientists better understand the regulation of gene expression in the body and how it leads to the complexity of cellular structures and functions. In addition, proteomics may lead to the development of breakthrough drugs for a variety of genetic disorders.

GENES AND OUR WORLD

Breakthroughs in decoding and manipulating the genetic information stored in DNA promise a world of benefits. These scientific advances already help to diagnose and treat disease, develop new medicines, bring criminals to justice, improve our food supply, and clean up the environment. At the same time, however, genetic technologies also present society with the potential for new and serious social or environmental problems. Many of the developments that worry critics of genetic technologies remain on the horizon, but the debate over their inevitable arrival is already in full swing.

Genes and the Environment

Humans have tampered with the genetic composition of other organisms for thousands of years. Most of this manipulation has been decidedly low-tech: domestication of animals and selective breeding of desirable food crops. The development and use of genetic engineering techniques has accelerated the pace at which humans can alter nature, creating some products that have unquestionable benefits and others that raise serious concerns.

Some scientists fear that genetic engineering techniques will damage genetic diversity. Domestication and selective breeding, which aim to produce many similar organisms with particular desirable traits, reduce the natural variation of genes within a species. Genetic engineering techniques take this effect a giant step further. Many crop species—including wheat, corn, tomatoes, and strawberries—have been manipulated to maximize yield, appearance, resistance to pests and chemicals, hardiness, and other commercially valuable traits.

Once attractive varieties have been developed, agricultural

techniques favour wide-scale planting of such genetically similar or identical stocks. The impact of this practice on biodiversity can be ominous because older plant varieties that carry diverse and useful alleles may be lost forever. The forfeited genetic material could leave a species vulnerable to annihilation from a single factor in the environment, such as an insect pest or an infectious virus.

Fortunately, efforts to combat decreasing biodiversity are under way. Farmers, gardeners, government agencies, and other interested parties have collaborated to create seed banks to maintain genetic diversity. These banks catalogue, store, and distribute the seeds of rare or endangered plants, enabling gardeners and farmers to continue cultivating rare plant varieties so that the unique genetic makeup of these plants does not disappear. In the same vein, zoos and other institutions breed endangered species of animals that may no longer be able to survive in their native habitats.

Other animal programs seek to preserve or enhance the genetic variation within certain endangered animal populations. For example, all cheetahs are almost identical genetically, most likely due to their near extinction about 12,000 years ago. Inbreeding among the few remaining individuals has resulted in a loss of genetic diversity in modern cheetahs that may have affected the cheetah's immune system, leaving the animal vulnerable to disease. Scientists hope to use genetic engineering techniques to introduce new genes into the cheetah population to increase the genetic diversity of the species.

The broad use of genetic engineering techniques in agriculture has raised other concerns beyond issues of biodiversity. From a consumer point of view, for instance, new technologies may have compromised a food's taste or nutritional value in exchange for a plant or animal that can be grown faster or at less cost. In addition, some critics question the safety of genetically engineered foods. They fear that plants or animals that have received new genes will produce proteins that would not be present in nonmanipulated organisms.

Such changes could have a serious effect: causing allergies or toxicity in humans who eat these foods, for example, or disrupting a plant's production of key nutrients. Though there has been much speculation about the potential health risks of genetically engineered foods, rigorous scientific investigation into the effects of these foods on humans is just beginning.

The potential environmental impact of genetically engineered agriculture is equally controversial. Critics fear that transgenic organisms, which contain DNA from other species, could give rise to populations of genetically altered life forms that could cause disease, displace native species, or otherwise harm delicate ecosystems. Consider the example of a genetically engineered form of oilseed rape, the plant that yields canola oil.

The altered form, which has been grown commercially in the United States since 1993, contains inserted genes that increase its resistance to herbicides (weed-killing chemicals). The altered plants can grow unabated even when a farmer sprays a field with enough herbicide to kill troublesome weeds. Yet a curious problem has emerged: The transgenic rape can interbreed with weedy relatives that grow nearby. Some scientists fear that this interbreeding could create weed varieties that also have a genetic resistance to known herbicides. The biological and economic impact of such a development could be enormous.

Genes and Society

Advances in genetic technologies allow scientists to take an unprecedented glimpse into the genetic makeup of every person. The information derived from this testing can serve many valuable purposes: It can save lives, assist couples trying to decide whether or not to have children, and help law-enforcement officials solve a crime. Yet breakthroughs in genetic testing also raise some troubling social concerns about privacy and discrimination.

For example, if an individual's genetic information becomes widely available, it could give health insurers cause to deny coverage to people with certain risk factors or encourage employers to reject certain high-risk job applicants. Furthermore, many genetically linked problems are more common among certain racial and ethnic groups—for example, the BRCA1 breast cancer allele is more common in Ashkenazi Jews, and the blood disorder sickle-cell anemia is more prevalent among blacks of African ancestry. Many minority groups fear that the expansion of genetic testing could create whole new avenues of discrimination.

Of particular concern are genetic tests that shed light on traits such as personality, intelligence, and mental health or potential abilities. Genetic tests that indicate a person is unlikely to get along with other people could be used to limit a person's professional advancement. In other cases, tests that identify a genetic risk of heart failure could discourage a person from competing in sports.

New technologies that allow the manipulation of genes have raised even more disturbing possibilities. Gene therapy advances, which allow scientists to replace defective genes with normal alleles, give people with typically fatal diseases new hope for healthy lives. To date, gene therapy has focused on manipulating the genetic material in body cells other than gametes, so the changes will not be passed on to future generations. However, the application of gene therapy techniques to gametes—the cells involved in reproduction—seems inevitable.

Such manipulation might help prevent the transmission of disease from one generation to another, but it could also produce unforeseen problems with long-lasting consequences. For instance, many people worry that new genetic techniques could be used to alter or encourage traits now viewed as part of normal human variability, such as shortness or baldness. At various times in the past century, people have advocated efforts to improve the human condition by promoting the perpetuation of certain genes.

This concept, known as eugenics, typically involves encouraging people with "positive" genes to reproduce and discouraging those with "inferior" genes from having offspring. Many people fear that new genetic technologies used to manipulate the human genome could give people previously unattainable methods to resort to extreme forms of eugenics.

Advances in genetic technologies have turned some genes into valuable commercial commodities, spawning a host of controversial questions. Who owns a genetically altered organism or the genes it contains? Is it right to patent the use of a naturally occurring gene? Some people feel that genetic material should not be owned or used for profit. Costa Rica has enacted laws to prevent foreign companies from patenting and then profiting from genes of native Costa Rican plant and animal species. Balancing the need to limit patents on genes are concerns that the profit motive of companies must be protected to maintain incentives to make new discoveries for medical products. The citizens of Iceland, for example, are cooperating with a biotechnology company in a study of the genetic makeup of the Icelandic people. The information compiled will be used to learn about genetic diseases.

HISTORY

Humans have had some understanding of heredity since prehistoric times, observing how similar traits pass from parent to offspring and noting that differences arise with each generation. Most of the mechanisms of heredity, however, were shrouded in mystery until early in the 20th century. Since that time, the rate of discovery has reached a feverish pace, enabling the advancement of modern molecular biology and the current Human Genome Project.

Early Views of Heredity

In ancient times, people understood some basic rules of heredity and used this knowledge to breed domestic animals and crops. By about 5000 BC, for example, people in different parts of the world had begun applying selective breeding techniques to grow new plant varieties, including types of wheat, maize, rice, and date palms, that had never existed in the wild.

Ancient people understood that the rules of inheritance

also applied to humans. The ancient Greeks were particularly interested in human heredity and evolution. Greek scientists and philosophers hotly debated whether a male or female parent contributed more to an offspring. In the 4th century BC, Aristotle speculated that acquired characteristics, such as a scar that was incurred during life, could be passed on to offspring.

He also believed in a widely held theory known as pangenesis. This theory proposed that particles in the body, called gemmules, reside in the limbs and organs. The gemmules become imprinted with any changes acquired by the body, such as muscle development from exercise. The gemmules then move to the reproductive cells and transfer information about the body's alterations to these cells. The reproductive cells transmit the acquired traits to offspring through particles called pangenes.

The theories about the inheritance of acquired characteristics and pangenesis persisted until the middle of the 19th century. French zoologist Jean-Baptiste Lamarck formalized the theory of acquired characteristics in his treatise *Philosophie Zoologique* (1809). Lamarck proposed that organisms evolve by responding to changes in their environment. When organisms undergo a change in order to adjust to their environment, that change acts as a trait that can be passed on to offspring.

INFLUENCES OF DARWIN AND MENDEL

A surprising supporter of pangenesis was the British naturalist Charles Robert Darwin, who believed that the theory accounted for the process of heredity and the wide variety of traits seen among offspring.

Despite his mistaken belief in pangenesis, Darwin nonetheless had an enormous impact on human understanding of heredity. During his years of extensive worldwide travel, Darwin collected many observations of how related species adapt to their local environments. Darwin and British naturalist Alfred Wallace independently formulated the theory of natural selection, which holds that members of a given species born with more favorable characteristics to deal with their environment would be most likely to survive to pass on these traits to the next generation. This important theory was popularized by Darwin's publication *On the Origin of Species* (1859). The book was an immediate sensation, but it raised many questions. Foremost among these was the mystery of how organisms could appear with modified or entirely new traits.

At roughly the same time that Darwin published his natural selection theories, the answer to many questions about the mechanisms of heredity were being unraveled by Gregor Mendel, a reclusive Austrian monk. Mendel conducted a long series of experiments on pea plants during the 1850s and 1860s. Mendel crossbred plants that expressed differing traits, such as height and flower colour.

His conclusions from these experiments helped him formulate a comprehensive theory of how such traits pass from one generation to another. In his studies, Mendel recognized that characteristics were inherited as discrete units, and that each of these was inherited independently of the others. He speculated that each parent has pairs of these units but passes only one to an offspring. He also noted that certain forms of one trait were always dominant over others. Today the units that Mendel described are known as genes.

EMERGENCE OF THE SCIENCE OF GENETICS

Mendel published his findings in 1866, but they went largely unnoticed for more than three decades. In the year 1900, however, Dutch botanist Hugo Marie de Vries, German botanist Karl Correns, and Austrian botanist Erich Tschermak independently rediscovered the monk's works and verified his conclusions.

Advances in cytology, the science of the structure and function of cells, enabled scientists to more deeply appreciate Mendel's work. In 1902 American biologist Walter S. Sutton and German cell biologist Theodor Boveri separately noted the parallels between Mendel's units and chromosomes. The demonstration of the chromosomal basis of inheritance gave rise to the modern science of genetics. The term *genetics* itself was coined in 1905 by British biologist William Bateson. The terms *gene* and *genotype* were contributed in 1909 by German scientist Wilhelm Johannsen.

In 1905 American biologists Edmund B. Wilson and Nettie Stevens independently discovered and identified the sex chromosomes. Wilson discovered the X chromosome in a butterfly, and Stevens discovered the Y chromosome in a beetle. The discoveries of the X and Y chromosomes helped scientists begin to unravel new patterns of inheritance. Foremost among this research was the work of American biologist Thomas Hunt Morgan on fruit flies. In 1910 Morgan identified the first proof of a sex-linked trait, an eye-colour characteristic that resides on the X chromosome of fruit flies. With this finding, Morgan became the first scientist to pin down the location of a gene to a specific chromosome. Morgan was also the first to explain the implications of linkage, unusual patterns of inheritance that occur when multiple genes found on the same chromosome are inherited together.

A student of Morgan's, American biologist Alfred Sturtevant, found early evidence of the mechanisms of crossing over, the phenomenon in which chromosomes interchange genes. More definitive proof emerged in the 1930s with work by American geneticists Harriet Creighton and Barbara McClintock. The pair demonstrated gene recombination with experiments on seed colour in corn. McClintock later gained notice for her work on transposable elements, large genetic segments that move within a chromosome or even between chromosomes. Her research into these elements, commonly known as jumping genes, earned McClintock the 1983 Nobel Prize in physiology or medicine.

BREAKTHROUGHS IN DNA STUDIES

While cytologists and geneticists were studying the properties and location of genes on chromosomes, other scientists focused their studies on the composition of genes. In 1928 British microbiologist Frederick Griffith ran a series of experiments on two strains of bacteria, one that kills mice and another that is harmless to them. When Griffith injected mice with killed cells of the virulent bacteria, all of the mice survived. But in a second trial, when Griffith injected a combined cocktail of dead virulent bacteria and live "harmless" bacteria, the mice all died. He concluded that something in the dead virulent cells "transformed" the hereditary material of normally harmless bacteria so that they became killers. Most scientists at the time theorized that the transforming factor was composed of a protein.

The real identity of the transforming factor in this experiment was not identified until 1944, when American geneticists Oswald Avery, Colin MacLeod, and Maclyn McCarty revisited Griffith's research. After isolating different molecular components from dead bacterial cells, Avery and his colleagues determined that DNA was the agent that transformed the live harmless bacteria into killers.

Despite a growing body of evidence about the function of DNA, many scientists were not ready to reject proteins as the hereditary material. The debate was largely quieted in 1952 by American geneticists Alfred Hershey and Martha Chase. Hershey and Chase showed that when a type of virus called a bacteriophage infects a bacterium, it is the virus's DNA—not protein—that enters the bacterium to cause infection. Their studies confirmed that DNA contained the virus's genetic information, which triggered viral replication within the bacteria.

The experiments of Hershey and Chase convinced most scientists that DNA was the molecule of heredity, but many questions about the structure and mechanisms of DNA remained. In the early 1950s researchers began to apply techniques of X-ray diffraction to learn about the basic structure of DNA. X-ray diffraction can determine molecular structures by measuring patterns of scattered X rays after they pass through a crystalline substance. British physical chemist Rosalind Franklin and British biophysicist Maurice Wilkins used X-ray diffraction to obtain DNA images of unprecedented clarity.

Yet the exact three-dimensional structure of DNA

remained unclear. The groundbreaking work of American biochemist James Watson and British biophysicist Francis Crick solved that mystery. In 1953 the two proposed a model of DNA that is still accepted today: A double helix molecule formed by two chains, each composed of alternating sugar and phosphate groups, connected by nitrogenous bases. Watson and Crick (along with Wilkins) were awarded the 1962 Nobel Prize in physiology or medicine for their discoveries.

Watson and Crick speculated that the structure of DNA provided some obvious clues about how the molecule could replicate itself. They proposed a replication model in which each strand of DNA serves as a template for making exact copies. This model of replication, called semi-conservative replication, was demonstrated in 1958 by American molecular biologists Matthew Meselson and Franklin Stahl. Their experiments demonstrated the mechanisms of replication by tracking DNA containing a heavy nitrogen isotope through a series of replications.

With DNA's structure and replication mechanisms largely solved, scientists turned their attention to identifying the genetic code—learning how a gene's nucleotide sequence determines what type of protein is made. In the late 1950s, South African geneticist Sydney Brenner and other scientists confirmed that RNA acted as an intermediary between DNA and protein production. Researchers still were uncertain how the sequence of nucleotides in DNA corresponded to the production of specific amino acids. In 1961 Crick and Brenner determined that groups of three nucleotides, now known as codons, code for the 20 amino acids that form the foundation of proteins.

The exact relationship between codons and amino acids was clarified after several important discoveries. American biochemists Marshall Nirenberg and J. Heinrich Matthaei synthesized repeated nucleotide sequences that led to the production of repeated single amino acids. They identified how certain codon combinations code for a specific amino acid. A process developed by American geneticist Har Gobind

Khorana helped scientists create a "dictionary" of codons that defined specific amino acids, thus resolving the remaining ambiguities in the genetic code. Only 12 years after the structure of DNA was deduced, the genetic code was solved.

Learning to Manipulate DNA

After scientists had unraveled the structure and replication mechanisms of DNA, many felt that the major discoveries of genetic research were resolved. They predicted that the only task left in genetics was to sort out the molecular details of how genes work. But in the process of studying gene function, researchers developed powerful new molecular techniques, enabling them to analyse and manipulate genes with a speed and precision never before possible.

A number of discoveries made during the 1960s and 1970s shed light on how distinct fragments of DNA could be isolated. The work of Swiss molecular biologist Werner Arber focused on specialized enzymes that digest, or "restrict," the DNA of viruses infecting bacteria. These enzymes were subsequently dubbed restriction enzymes. In the following decade, scientists learned that restriction enzymes could also act like molecular scissors to cut DNA. In 1970 American molecular biologist Hamilton Smith and colleagues determined that restriction enzymes could cleave DNA molecules at precise and predictable locations. Hamilton concluded that the enzymes were able to recognize specific nucleotide sequences.

Scientists quickly realized that restriction enzymes could be used in the laboratory to manipulate DNA. In 1973 American biochemist Herb Boyer used restriction enzymes to produce a DNA molecule with genetic material from two different sources. This splicing technique is now known as recombinant DNA. Boyer inserted foreign genes into plasmids and observed that the plasmids could replicate to make many copies of the inserted genes.

In subsequent experiments, Boyer, American biochemist Stanley Cohen, and other researchers demonstrated that inserting a recombinant DNA molecule into a host bacteria cell would lead to extremely rapid replication and the production of many identical copies of the recombinant DNA. This process, known as cloning, gave scientists the power to make many copies of desired DNA for molecular study.

The speed and efficiency of DNA cloning were vastly improved in the 1980s with the invention of polymerase chain reaction (PCR). Developed by American biochemist Kary Mullis, PCR enables scientists to produce large amounts of DNA sequences in a test tube. In a matter of hours, the process can produce millions of cloned DNA molecules. Yet all of the advances in isolating and replicating DNA would not be possible or be of much use if researchers could not determine the nucleotide sequence of genetic material. In the late 1970s and early 1980s, British biochemist Frederick Sanger and his associates developed DNA sequencing techniques. Sanger's methods, which used special compounds called dideoxy nucleotides, rapidly yielded the exact nucleotide sequence of a desired sample. With the use of automated equipment, the new techniques transformed genetic sequencing into a speedy, routine laboratory procedure.

Many of the new techniques for isolating, sequencing, and replicating DNA have been put to practical use through the field of genetic engineering. The Human Genome Project and the new field of proteomics have both benefited from continuing technical advances and have accelerated the development of new genetic technologies. Modern genetics is poised to radically change the practice of medicine and the biotechnology industry.

THE TOOLS OF GENETIC ENGINEERING

Cutting and Measuring DNA

The first step in gene cloning is cutting DNA into appropriately sized pieces. The size of DNA molecules is measured by subjecting the DNA to a technique called "electrophoresis". An electrophoretic apparatus gives DNA an electrical shove (DNA is negatively charged because of its phosphate groups. The electrophoresis apparatus puts the DNA in an electric field, and it moves toward the positive pole), forcing it to move through a porous gel. Two types of gels are commonly used.

Agarose Gels

The most popular variant of electrophoresis is carried out with agarose gels. In the laboratory, suspensions of agarose in buffer are heated to boiling (often in microwave ovens), and the hot solution is poured on plastic or glass plates to a height of a few millimeters or so. Agarose is a derivative of agar — an edible polysaccharide extracted from seaweed.

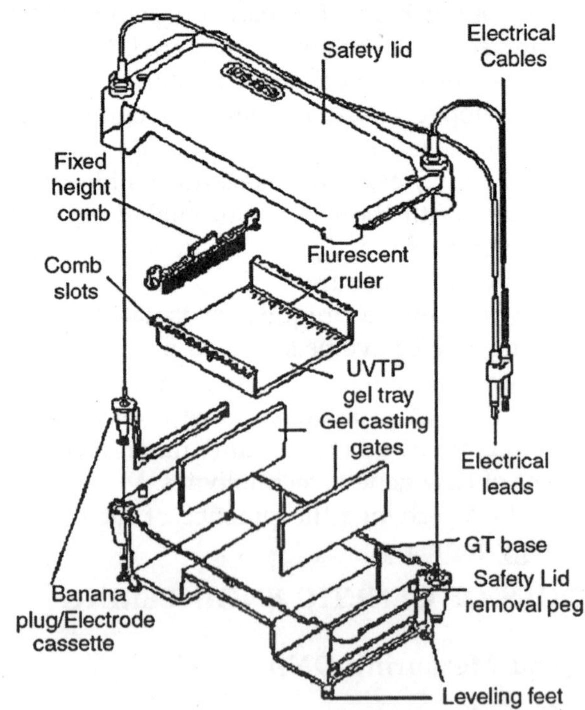

Fig. Electrophoresis

When cooled from heated solutions, solutions of agarose form a nearly transparent gel that looks like a slab of gelatin dessert. After the gel sets, DNA solutions are placed in little depressions formed by leaving a comb in the gel. The gel is then flooded with a weak salt solution, and a voltage is applied.

Since DNA is negatively charged (due to its phosphate groups), it moves toward the positive pole. Large DNA molecules advance slowly in the gel because they are impeded by the gel matrix. Smaller ones encounter fewer barriers and move more quickly. By plotting the distance migrated against the reciprocal of size (in base pairs) of a group of DNA standards, a fairly straight line can be obtained. From these data, the length of unknowns can be easily calculated. For ease of comparison, the standards are usually run in the same gel beside the unknowns.

Gels of differing porosities can be made by adjusting the concentration of agarose. With 2 and 3% solutions, double-stranded DNA's as small as 50 or 100 base pairs can be resolved. More dilute gels can resolve fragments as large as about 30,000 base pairs (30 kilobase pairs or 30kb).

Polyacrylamide Gels

When very small DNA molecules need to be analyzed or when high-resolution separations are required, electrophoresis is carried out in polyacrylamide gels. They, like the gels made from agarose, are water white. Unlike agarose, polyacrylamide gels are polymers of acrylamide, a small synthetic organic compound.

Polyacrylamide gel electrophoresis (PAGE) is most often used to separate fragments of DNA that are between 6 and a few thousand base pairs in length. The technique is used mostly in DNA sequencing, a topic that we'll come back to later in the course.

Detecting DNA

DNA doesn't have any colour; it only absorbs light in the ultraviolet range. How is it visualized in a clear gel after electrophoresis? DNA can be detected in both agarose and polyacrylamide gels after staining with various dyes, incuding ethidium bromide, a dye that forms a fluorescent complex upon binding to DNA.

Usually, the ethidium bromide is added to the gel before electrophoresis. After the run, the gel is examined and often photographed under an ultraviolet light. Yellow-orange zones (bands) of fluorescence appear (similar to those in the picture at the right), indicating migration of discrete pieces of DNA. Other agents — like methylene blue or silver stains — can also be used to visualize DNA, but they are either less sensitive or less convient, and aren't used as often. By the way, many dyes that bind to DNA are mutagenic; i.e, they cause mutations.

Separating Large Fragments of DNA

Contrary to what we might expect, very large fragments of DNA readily move in agarose gels when an electrical field is applied. But they don't separate according to size. In fact, very big pieces of DNA all migrate at about the same velocity regardless of whether they are 20kb in length or 2000kb.

PFE

Since many techniques, particulary recently derived ones, depend on working with very large DNA fragments, even pieces the size of some chromosomes, several newly developed electrophoretic techniques have been developed. In one of these, called pulsed field electrophoresis (PFE), DNA molecules are analyzed on agarose gels but with a modified apparatus. Instead of having only two sets of electrodes situated at opposite ends of the electrophoresis device as in the picture shown above, the pulsed field apparatus bears four sets of electrodes.

Two are designated A- and A+ and the other two are called B- and B+. The two A's and the two B's are set at an angle of 120° with respect to one another. During electrophoresis, the DNA molecules are subjected to alternating bursts of current from the two A and B pairs, thereby pulling the DNA alternately to the right and to the left as it advances (from south to north in this case) through the gel. The DNA seems to reorient itself each time the current is switched, and the time that it takes to turn itself around is dependent on its length.

The result is that very large pieces of DNA can be separated from another on the basis of their size. The technique also allows one to estimate the length of an unknown when run beside standards of known size. One of the most impressive accomplishments of this technique is the separation of the 16 chromosomes of yeast in a single electrophoretic run.

FIE

Another procedure, field inversion electrophoresis, is also effective at separating large molecules of DNA. In it, the electrodes are set up as in ordinary electrophoresis, but the current is switched so that the DNA first moves forward and then backward.

Of course, if the switching were to be done in equal time intervals — say, 1 second forward and 1 second back — the DNA wouldn't move at all. To get it to move at all, electrophoresis is conducted forward for a longer time than backward (or at a greater voltage forward versus backward).

One might expect that moving ahead three steps and back two would be equivalent to moving forward one step at a time, but that's not what happens. Apparently the DNA reorients itself during the switching cycles, just as it does in pulsed field electrophoresis.

Again, the ability to change directions seems to be dependent on size. This allows different lengths of DNA to be separated from one another.

Restriction Endonucleases

One critically important advance that has greatly stimulated the rapid progress in molecular biology and genetic engineering was the discovery of a set of enzymes that are capable of cutting DNA at defined sequences. These enzymes are found in a variety of microorganisms and are called restriction endonucleases, or more simply, restriction enzymes.

The first specific restriction enzyme was discovered by Hamilton Smith in 1970, an accomplishment for which he (and two others) were awarded the Nobel Prize. Since then, some 3,059 similar enzymes have been reported (as of April 6, 1999), many of which have different specificities.

Nomenclature

Smith and another Nobel laureate, Daniel Nathans, devised a nomenclature for these enzymes. In brief, the name of each restriction enzyme derives from the organism from which it is isolated. The first letter of the genus name plus the first two letters of the species name form the first three letters of the restriction enzyme's name.

If necessary, a letter indicating strain designation is added, and finally a number is appended that stands for the order in which the enzyme was discovered in each organism. For example, BamHI is the name of an enzyme that is isolated from the bacterium *Bacillus amyloliquifaciens,* strain H, and it was, presumably, the first restriction endonuclease identified from that source. The first three letters of the name should be italicized, but because italicized letters are often hard to read on a computer, I have left them in plain text throughout these notes.

What they Do

The restriction enzymes owe their usefulness to the fact that they bind to DNA at specific DNA sequences, four to eight nucleotides in size, called recognition sites. Once bound, restriction enzymes cut the DNA at or near this site. With a little thought, it should be clear that an enzyme that has a six base pair recognition site will, on the average, produce larger pieces of DNA than one that recognizes a four base site.

Expressed quantitatively, the approximate size of the fragments produced by a particular enzyme, given that it is cutting a DNA containing an equal proportion of all four nucleotides, can be calculated from the formula:

$$\text{Average size of fragment} = 4^N$$

where N is the number of bases that the enzyme recognizes. Hence, a four-cutter (the shorthand name for an enzyme that recognizes a site containing four base pairs) is expected to cleave random DNA into fragments of about 4^4 (256) base pairs while an enzyme with a recognition site of six bases will produce pieces (on the average) of about 4^6 (4096) base pairs.

PROPERTIES OF RESTRICTION ENZYMES

Ends

Another interesting property of restriction enzymes is that while they often recognize a symmetrical site, they do not always cut at the axis of symmetry. For instance, the enzyme EcoRI (from *Escherichia coli* strain RY13 and pronounced "echo are one"), recognizes the site GAATTC and cuts the DNA between the G's and the A's in a manner depicted below.

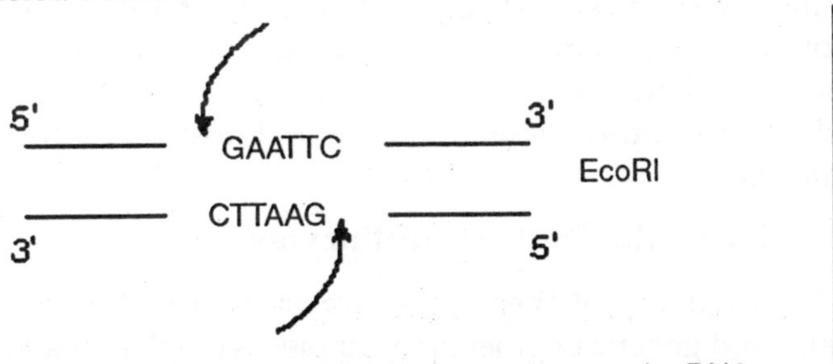

The arrows indicate where the enzyme cuts the DNA

Similarly, the enzyme BglII (from the microorganisms, *Bacillus globiggi* and universally and irreverently pronounced BAGEL TWO) recognizes the sequence AGATCT and cuts between the first A and G residues.

In fact, the four nucleotide single-stranded ends are the same for both BglII and BamHI. Moreover, there are at least two other six cutting enzymes that have been discovered that leave the same four nucleotide overhang: BclI and XhoII. These overhanging ends are very useful because — under the proper conditions — they may base pair with each other. In fact, because of their affinity for one another, they are often called cohesive or sticky ends.

Moreover, if molecules with these ends are treated with the appropriate enzyme — DNA ligase — their phosphodiester bonds may be rejoined (ligated). When two ends that originate from digestion by a single enzyme are ligated, the resulting molecule can be cut by the same enzyme again. But if the ends of a DNA molecule that originated with a BamHI cut and a

BglII cut are joined together, the new sequence will not be cut with either enzyme. All restriction endonucleases do not generate 5' single-strand overhangs. In fact, some don't even produce an overhang at all. Several enzymes — like SacI — produce 3' single-stranded sticky ends.

And some enzymes — like PvuII — cut at the axis of symmetry, leaving perfectly aligned ends. DNA molecules without overhangs are said to have blunt ends.In addition there are restriction enzymes that cleave DNA some distance away from the sequence that they recognize. For example the enzyme HgaI makes staggered cuts that lie 5 and 10 nucleotides away from a 5 base pair sequence, GACGC. This leaves 5' overhanging ends, but, in contrast to the enzymes described above, these will be different almost every time the enzyme cuts.

The Utility of the Restriction Enzymes

The discovery of these many restriction endonucleases have allowed genetic engineers to cut pieces of DNA at specific sites and into defined sizes. The result has been that a scientist can work with a collection of molecules all of the same size and with ends of known sequence. Restriction enzymes have proved to be valuable analytical and diagnostic tools as well.

RESTRICTION ENZYMES

Restriction enzymes are DNA-cutting enzymes found in bacteria (and harvested from them for use). Because they cut within the molecule, they are often called restriction endonucleases. A restriction enzyme recognizes and cuts DNA only at a particular sequence of nucleotides. For example, the bacterium Hemophilus aegypticus produces an enzyme named HaeIII that cuts DNA wherever it encounters the sequence

5'GGCC3'

3'CCGG5'

The cut is made between the adjacent G and C. This particular sequence occurs at 11 places in the circular DNA molecule of the virus phiX174. Thus treatment of this DNA with the enzyme produces 11 fragments, each with a precise length and nucleotide sequence. These fragments can be separated from one another and the sequence of each determined.

HaeIII and AluI cut straight across the double helix producing "blunt" ends. However, many restriction enzymes cut in an offset fashion. The ends of the cut have an overhanging piece of single-stranded DNA. These are called "sticky ends" because they are able to form base pairs with any DNA molecule that contains the complementary sticky end. Any other source of DNA treated with the same enzyme will produce such molecules.

Mixed together, these molecules can join with each other by the base pairing between their sticky ends. The union can be made permanent by another enzyme, DNA ligase, that forms covalent bonds along the backbone of each strand. The result is a molecule of recombinant DNA (rDNA).

The ability to produce recombinant DNA molecules has not only revolutionized the study of genetics, but has laid the foundation for much of the biotechnology industry. The availability of human insulin (for diabetics), human factor VIII (for males with hemophilia A), and other proteins used in human therapy all were made possible by recombinant DNA.

RECOMBINANT DNA AND GENE CLONING

Recombinant DNA is DNA that has been created artificially. DNA from two or more sources is incorporated into a single recombinant molecule.

Making Recombinant DNA (rDNA)

- Treat DNA from both sources with the same restriction endonuclease (BamHI in this case).
- BamHI cuts the same site on both molecules
  5' GGATCC 3'
  3' CCTAGG 5'
- The ends of the cut have an overhanging piece of single-stranded DNA.
- These are called "sticky ends" because they are able

to base pair with any DNA molecule containing the complementary sticky end.

- In this case, both DNA preparations have complementary sticky ends and thus can pair with each other when mixed.
- DNA ligase covalently links the two into a molecule of recombinant DNA.

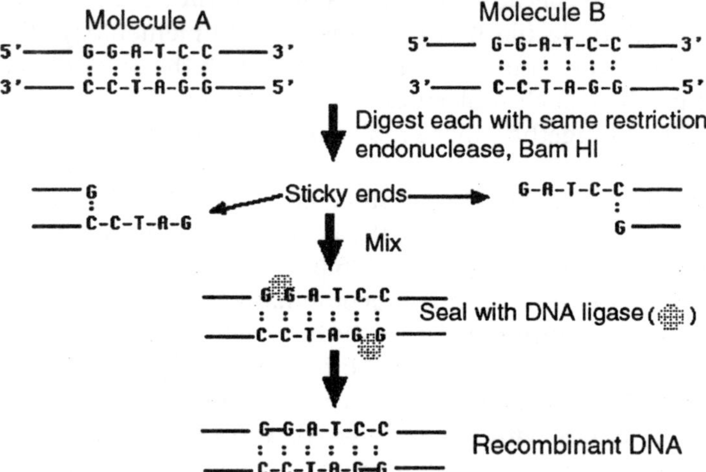

To be useful, the recombinant molecule must be replicated many times to provide material for analysis, sequencing, etc. Producing many identical copies of the same recombinant molecule is called cloning. Cloning can be done in vitro, by a process called the polymerase chain reaction (PCR). Here, however, we shall examine how cloning is done in vivo.

Cloning in vivo can be done in

- Unicellular microbes like E. coli
- Unicellular eukaryotes like yeast and
- in mammalian cells grown in tissue culture.

In every case, the recombinant DNA must be taken up by the cell in a form in which it can be replicated and expressed. This is achieved by incorporating the DNA in a vector. A number of viruses (both bacterial and of mammalian cells) can serve as vectors. But here let us examine an example of cloning using E. coli as the host and a plasmid as the vector.

PLASMID

A plasmid is an extra-chromosomal DNA molecule separate from the chromosomal DNA which is capable of replicating independently of the chromosomal DNA. In many cases, it is circular and double-stranded. Plasmids usually occur naturally in bacteria, but are sometimes found in eukaryotic organisms (e.g., the 2-micrometre-ring in Saccharomyces cerevisiae).

Plasmid size varies from 1 to over 200 kilobase pairs (kbp). The number of identical plasmids within a single cell can be zero, one, or even thousands under some circumstances.

Plasmids can be considered to be part of the mobilome, since they are often associated with conjugation, a mechanism of horizontal gene transfer.The term plasmid was first introduced by the American molecular biologist Joshua Lederberg in 1952.

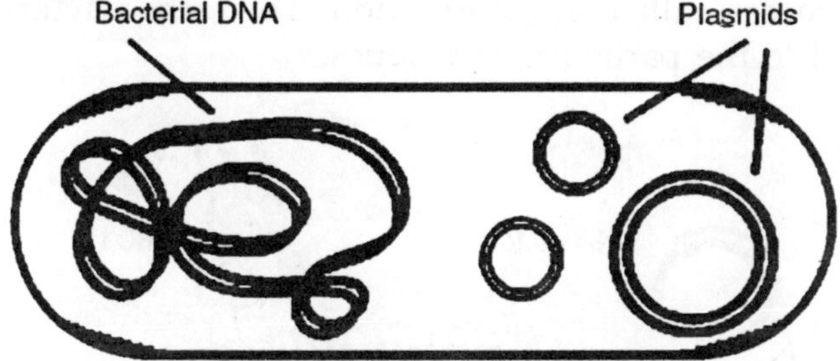

Fig: Plasmids Enclosed Showing Chromosomal DNA and Plasmids

Plasmids can be considered to be independent life-forms similar to viruses, since both are capable of autonomous replication in suitable (host) environments. However the plasmid-host relationship tends to be more symbiotic than parasitic (although this can also occur for viruses, for example with Endoviruses) since plasmids can endow their hosts with useful packages of DNA to assist mutual survival in times of severe stress.

For example, plasmids can convey antibiotic resistance to host bacteria, who may then survive along with their life-saving guests who are carried along into future host generations.

Vectors

There are two types of plasmid integration into a host bacteria: Non-integrating plasmids replicate as in the top instance; whereas episomes, the lower example, integrate into the host chromosome.

Plasmids used in genetic engineering are called vectors. Plasmids serve as important tools in genetics and biotechnology labs, where they are commonly used to multiply (make many copies of) or express particular genes. Many plasmids are commercially available for such uses.

The gene to be replicated is inserted into copies of a plasmid containing genes that make cells resistant to particular antibiotics and a multiple cloning site (MCS, or polylinker), which is a short region containing several commonly used restriction sites allowing the easy insertion of DNA fragments at this location. Next, the plasmids are inserted into bacteria by a process called transformation. Then, the bacteria are exposed to the particular antibiotics.

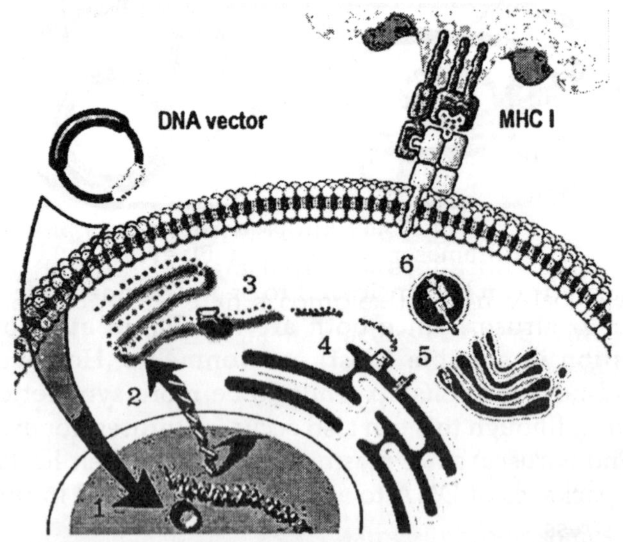

Fig. Plasmid Vector

Only bacteria which take up copies of the plasmid survive the antibiotic, since the plasmid makes them resistant. In particular, the protecting genes are expressed (used to make a protein) and the expressed protein breaks down the antibiotics. In this way the antibiotics act as a filter to select only the modified bacteria. Now these bacteria can be grown in large amounts, harvested and lysed (often using the alkaline lysis method) to isolate the plasmid of interest.

Another major use of plasmids is to make large amounts of proteins. In this case, researchers grow bacteria containing a plasmid harboring the gene of interest. Just as the bacteria produces proteins to confer its antibiotic resistance, it can also be induced to produce large amounts of proteins from the inserted gene. This is a cheap and easy way of mass-producing a gene or the protein it then codes for, for example, insulin or even antibiotics.

However, a plasmid can only contain inserts of about 1-10 kbp. To clone longer lengths of DNA, lambda phage with lysogeny genes deleted, cosmids, bacterial artificial chromosomes or yeast artificial chromosomes could be used.

TYPES

One way of grouping plasmids is by their ability to transfer to other bacteria. Conjugative plasmids contain so-called tra-genes, which perform the complex process of conjugation, the transfer of plasmids to another bacterium. Non-conjugative plasmids are incapable of initiating conjugation, hence they can only be transferred with the assistance of conjugative plasmids, by 'accident'.

An intermediate class of plasmids are mobilizable, and carry only a subset of the genes required for transfer. They can 'parasitize' a conjugative plasmid, transferring at high frequency only in its presence. Plasmids are now being used to manipulate DNA and may possibly be a tool for curing many diseases.

It is possible for plasmids of different types to coexist in a single cell. Seven different plasmids have been found in E. coli. But related plasmids are often incompatible, in the sense that only one of them survives in the cell line, due to the regulation of vital plasmid functions. Therefore, plasmids can be assigned into compatibility groups. Another way to classify plasmids is by function.

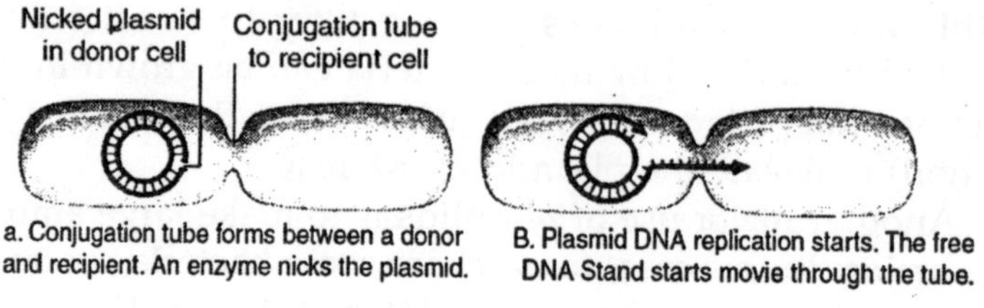

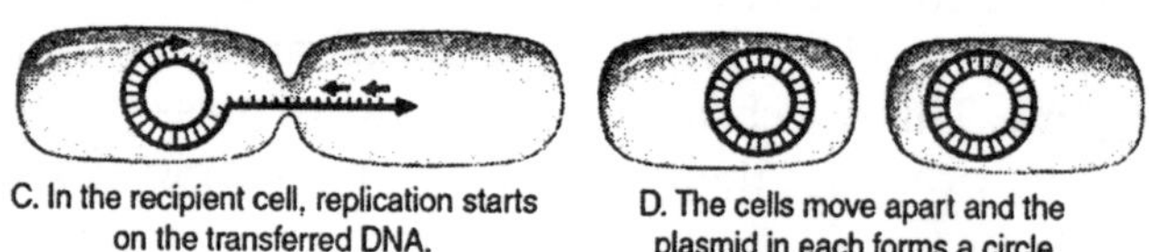

Fig: Bacterial conjugation

There are five main classes:

- Fertility-F-plasmids, which contain tra-genes. They are capable of conjugation.
- Resistance-(R)plasmids, which contain genes that can build a resistance against antibiotics or poisons. Historically known as R-factors, before the nature of plasmids was understood.
- Col-plasmids, which contain genes that code for (determine the production of) bacteriocins, proteins that can kill other bacteria.
- Degradative plasmids, which enable the digestion of unusual substances, e.g., toluene or salicylic acid.
- Virulence plasmids, which turn the bacterium into a pathogen.

Plasmids can belong to more than one of these functional groups. Plasmids that exist only as one or a few copies in each bacterium are, upon cell division, in danger of being lost in one of the segregating bacteria. Such single-copy plasmids have systems which attempt to actively distribute a copy to both daughter cells. Some plasmids include an addiction system or "postsegregational killing system (PSK)", such as the hok/sok (host killing/suppressor of killing) system of plasmid R1 in Escherichia coli. They produce both a long-lived poison and a short-lived antidote. Daughter cells that retain a copy of the plasmid survive, while a daughter cell that fails to inherit the plasmid dies or suffers a reduced growth-rate because of the lingering poison from the parent cell.

Plasmid DNA Extraction

As alluded to above, plasmids are often used to purify a specific sequence, since they can easily be purified away from the rest of the genome. For their use as vectors, and for molecular cloning, plasmids often need to be isolated. There are several methods to isolate plasmid DNA from bacteria, the archetypes of which are the miniprep and the maxiprep/ bulkprep. The former can be used to quickly find out whether the plasmid is correct in any of several bacterial clones. The yield is a small amount of impure plasmid DNA, which is sufficient for analysis by restriction digest and for some cloning techniques. In the latter, much larger volumes of bacterial suspension are grown from which a maxi-prep can be performed. Essentially this is a scaled-up miniprep followed by additional purification. This results in relatively large amounts (several micrograms) of very pure plasmid DNA

DNA LIGASES

DNA ligases are vital enzymes required for important cellular processes such as DNA replication, repair of damaged DNA and recombination. The enzyme mediates the formation of phosphodiester bonds between adjacent 3'-OH and 5'-phosphate termini, thereby joining the nicks in double stranded DNA. Ligases can be classified into two groups depending on their requirement for ATP or NAD+ as the cofactor. All eukaryotic and virally encoded enzymes are ATP-dependent, whereas most prokaryotic enzymes require NAD^+ for their activity.

ATP-Dependent DNA Ligase

DNA ligase from bacteriophage T7 is a monomer with a

molecular weight of 41 kDa. Crystal structures of the apo form of this enzyme and its complex with ATP have been determined at 2.7 Å and 2.6 Å respectively.

The structure consists of two distinct domains, a larger N-terminal domain and a C-terminal domain. The ATP-binding site is situated in the N-terminal domain in a pocket beneath one of the beta-sheets. This pocket is lined by a number of motifs that are conserved across a wide family of nucleotidyltransferases. The structure of the C-terminal domain is remarkably similar to the oligonucleotide binding fold, observed in a number of proteins. The DNA-binding site is proposed to be in a groove running between the two domains.

The structure has provided the basis for biochemical experiments. We have made the two domains separately and looked at their biochemical properties. The larger N-terminal domain is an active ligase but with greatly reduced activity. Both domains are able to bind to DNA. The N-terminal domain binds both single and double stranded DNA, while the smaller C-terminal domain is only able to bind double stranded DNA despite having a fold that is similar to single strand DNA binding proteins.

The affinity for nicked DNA comes from a combination of these two DNA-binding affinities at the active site. Interestingly, the N-terminal domain is poor at the adenylation reaction, but this activity is stimulated greatly by addition of the C-terminal domain.

Furthermore, we can demonstrate an association of the domains on gel filtration. These data suggest that a conformational change occurs during the adenylation reaction. This would be similar to that observed directly in our crystal structure of an mRNA capping enzyme, a group of enzymes that are closely related to ligases.

NAD-Dependent DNA Ligase

An NAD-dependent DNA ligase from *B.stearothermophilus*. The enzyme has been overexpressed and biochemical studies are in progress. Limited proteolysis has been used to investigate the domain structure of the enzyme. There are two domains, with the larger N-terminal domain retaining full self adenylation activity and the smaller C-terminal domain having all of the DNA-binding activity of the full length enzyme.

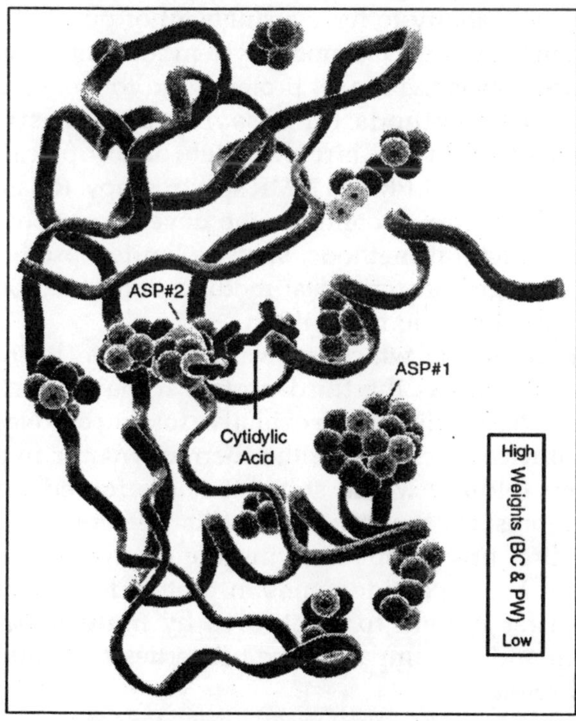

Fig: Structure of RNase A

The situation is therefore different to that we found in the ATP-dependent enzyme, where both domains contribute to the activities of the enzyme. Structural studies are underway with these domains. The structure of the large N-terminal domain has been solved and work is in progress to determine the structure of the small domain.

RIBONUCLEASE A

Ribonuclease A (RNase A) is an endonuclease that cleaves single-stranded RNA. Bovine pancreatic RNase A is one of the classic model systems of protein science.

HISTORY

The importance of bovine pancreatic RNase A was secured when the Armour & Co. (of hot dog fame) purified a *kilogram* of it, and gave 10 mg samples away free to any interested scientists. The ability to have a single lot of purified enzyme instantly made RNase *the* model system for protein studies.

RNase A was the model protein used to work out many spectroscopic methods for assaying protein structure, including absorbance, circular dichroism/optical rotary dispersion, Raman, EPR and NMR spectroscopy. RNase A was also the first model protein for the development of several chemical structural methods, such as limited proteolysis of disordered segments, chemical modification of exposed side chains, and antigenic recognition.

Ribonuclease-S, which is RNase A that has been treated with subtilisin, was the third protein to have its structure solved, in 1967.Studies of the oxidative folding of RNase A led Chris Anfinsen to enunciate the thermodynamic hypothesis of protein folding, which states that the folded form of a protein represents the minimum of its free energy.RNase A was the first protein for showing the effects of non-native isomers of X-Pro peptide bonds in protein folding.RNase A was the first protein to be studied by multiple sequence alignment and by comparing the properties of evolutionarily related proteins.

STRUCTURE AND PROPERTIES

Labeled ribbon diagram of bovine pancreatic ribonuclease A (PDB accesion code 7RSA). The backbone ribbon is colored from blue (N-terminus) to red (C-terminus). The side chains of the four disulfide-bonded cysteines are shown in yellow, with their sulfur atoms highlighted as small spheres. Residues important for catalysis are shown in magenta.

RNase A is a relatively small protein (124 residues, ~13.7 kDa). It can be characterized as a two-layer á + â protein that is folded in half to resemble a taco, with a deep cleft for binding the RNA substrate. The first layer is composed of three alpha helices from the N-terminal half of the protein. The second layer consist of three â-hairpins arranged in two â-sheets. The hairpins 61-74 and 105-124 form a four-stranded, antiparallel â-sheet that lies on helix 3 (residues 50-60). The longest â-hairpin 79-104 mates with a short â-strand (residues 42-45) to form a three-stranded, antiparallel â-sheet that lies on helix 2 (residues 24-34).

RNase A has four disulfide bonds in its native state: Cys26-Cys84, Cys58-110, Cys40-95 and Cys65-72. The first two (26-84 and 58-110) are essential for conformational folding; each joins an alpha helix of the first layer to a beta sheet of the second layer, forming a small hydrophobic core in its vicinity.

The latter two disulfide bonds (40-95 and 65-72) are less essential for folding; either one can be reduced (but not both) without affecting the native structure under physiological conditions. These disulfide bonds connect loop segments and are relatively exposed to solvent.

Interestingly, the 65-72 disulfide bond has an extraordinarily high propensity to form, significantly more than would be expected from its loop entropy, both as a peptide and in the full-length protein. This suggests that the 61-74 â-hairpin has a high propensity to fold conformationally.

RNase A is a basic protein (pI =8.63); its many positive charges are consistent with its binding to RNA (a poly-anion). More generally, RNase A is unusually polar or, rather, unusually lacking in hydrophobic groups, especially aliphatic ones. This may account for its need of four disulfide bonds to stabilize its structure.

The low hydrophobic content may also serve to reduce the physical repulsion between highly charged groups (its own and those of its substrate RNA) and regions of low dielectric constant (the nonpolar residues). The N-terminal á-helix of RNase A (residues 3-13) is connected to the rest of RNase A by a flexible linker (residues 16-23).

As shown by F. M. Richards, this linker may be cleaved by subtilisin between residues 20 and 21 without causing the N-terminal helix to dissociate from the rest of RNase A. The peptide-protein complex is called RNase S, the peptide (residues

1-20) is called the S-peptide and the remainder (residues 21-124) is called the S-protein.

The dissociation constant of the S-peptide for the S-protein is roughly 30 pM; this tight binding can be exploited for protein purification by attaching the S-peptide to the protein of interest and passing a mixture over an affinity column with bound S-protein.

The RNase S model system has also been used for studying protein folding by coupling folding and association. The S-peptide was the first peptide from a native protein shown to have (flickering) secondary structure in isolation.

Chapter 3

Bioinformatics

Before long, the DNA sequence of the complete human genome will have been determined. This achievement might seem an end in itself, but it is really only the beginning. Already a large number of bacterial genomes have been fully sequenced, an outstanding achievement in science that started 1995 with the completion of the bacterium Haemophilus influencae.

Bacterial genomes were to be followed by the first eukaryotic organism, the unicellular genetic model system Saccharomyces cerevisiae, more commonly known as baker's yeast. In December 1998, the first multicellular organism has been added to the list, the nematode Caenorhabditis elegans, which now provide us with information about unique functions in organisms of greater complexity. The sum of all this information is enormous and its potential in our understanding of life processes, if rightly explored, is far reaching and tremendous.

However, today with modern genome wide analytical technologies available, such as micro array transcript analysis or global protein analysis utilising electrophoresis and mass spectrometry, it is biochemical and molecular biology data in general, rather than DNA sequence data alone, that is accumulating at a phenomenal rate. In order to exploit this wealth of information a new field of science has arisen that fuses biology and medicine on one side with mathematics, statistics and computer science on the other side. This new field of science is known as bioinformatics.

All sequence data is compiled in large international

databases, soon to be followed by data collections on e.g. expression data, protein-protein interaction data, phenotypic data for mutants etc. Straightforward access to data over the Internet means that a wealth of information is available, literally at our fingertips. The topics covered within Bioinformatics range from retrieving and aligning DNA and protein sequences to predicting structure and function of gene products. We will here only briefly touch on the many facets of bioinformatics.

GENOME AND SEQUENCE ANALYSIS

Historically, bioinformatics as a concept was invented to describe the task of handling, presenting and analysing large amounts of sequence data. Today, due to intense efforts at a number of large research centres throughout the world, data can be rather easily accessed by anyone over the Internet and World Wide Web servers.

As a consequence, it is currently almost an everyday activity in most molecular biology labs to screen these sequence databases to find sequence homologues of a particular gene. This is not only to find homologues within a species but also to look for similar genes in other organisms, so called orthologous. The discovery of numerous such orthologous groups of genes provide excellent support for the power of using of model organisms. Sequence similarities is also used to cluster organisms according to their evolutionary relatedness and thus to create phylogenetic trees, an important tool in taxonomy.

In parallel to the DNA sequencing effort, determination of the location of genes on chromosomes is today performed in large scale projects for a number of organisms, which provide information that need to be efficiently handled and presented.

FROM SEQUENCE TO 3D STRUCTURAL PREDICTION

For most macromolecules their function is closely linked to the three-dimensional structure, maybe most apparent for proteins and some RNA molecules. Recent technical developments can now provide us with a more detailed view of how molecules are folded. The experimental determination of these 3D structures is, however, a costly and slow process. Novel procedures for predicting the molecular fold from the primary sequence data is thus urgently needed. Since the protein structure is ultimately carrying the information about the enzymatic active site or surface site for protein-protein interaction, knowledge about protein tertiary structure will in the future be of fundamental importance for the pharmaceutical industry.

ANALYSIS OF GENOME WIDE BIOMEDICAL DATA AND FUNCTIONAL GENOMICS

In the last couple of years the advent of biomedical large scale analysis tools have for ever changed the way scientists in biology and medicine will do research. These technologies make possible the simultaneous study of the expression of thousands of genes, either at the transcript or at the protein level, or the thousands of possible protein-protein interactions in a cell, or phenotypic analysis of thousands of mutants etc. All this data, regardless of type and format, has to be handled, presented and efficiently analysed.

This challenge is already being explored by statisticians for the clustering of e.g. similarly regulated genes. This clustering information is currently being evaluated as a potentially useful way of predicting function of functionally uncharacterised genes in the following up on the genomics projects, a research area called functional genomics. Ultimately, prediction of gene function will include a more complex procedure, i.e. the integrated analysis of many types of large scale molecular data into one tentative function for the studied gene.

This latter task will of course also utilise information gained by applying the above described sequence analysis. In addition, genes with similar expression profiles would possibly exhibit consensus sequence elements in their regulatory regions. Identifying these sequences by automated computer methods, which is more difficult than finding clear similarities between the encoded proteins, will be a great challenge that can provide extremely useful information.

MATHEMATICAL MODELLING OF LIFE PROCESSES

The vast amounts of data generated by the genome wide analytical technologies will not only have to be clustered, but also more importantly, interpreted in a physiological context. To be able to do so in a more sophisticated manner than is currently possible when handling thousands of information units, automated strategies have to be developed. This is a formidable task that incorporates modelling of all molecular processes in a cell at the molecular level.

Initially, this task will be approached by modelling of discrete parts of the cell's physiology, like metabolic fluxes or regulatory networks. However, the integration of all these will in many ways be the ultimate challenge for bioinformatics and an important part of the final goal of biomedical science in general - the complete molecular understanding of a living organism.

DATABASE BUILDING AND MANAGEMENT

Whatever type of information is being generated, analysed and finally interpreted, the data has to be presented to the scientific community by establishing Internet based World Wide Web servers.The presentation of this data can be rather challenging, and problems that arise extend from formalism of data submission to intelligent and clear ways of presentation. Database management is thus not only an engineering problem, but also provides a clear scientific challenge.

INTERNET

Internet, computer-based global information system. The Internet is composed of many interconnected computer networks. Each network may link tens, hundreds, or even thousands of computers, enabling them to share information and processing power. The Internet has made it possible for people all over the world to communicate with one another effectively and inexpensively. Unlike traditional broadcasting media, such as radio and television, the Internet does not have a centralized distribution system. Instead, an individual who has Internet access can communicate directly with anyone else on the Internet, post information for general consumption, retrieve information, use distant applications and services, or buy and sell products.

The Internet has brought new opportunities to government, business, and education. Governments use the Internet for internal communication, distribution of information, and automated tax processing. In addition to offering goods and services online to customers, businesses use the Internet to interact with other businesses. Many individuals use the Internet for communicating through electronic mail (e-mail), retrieving news, researching information, shopping, paying bills, banking, listening to music, watching movies, playing games, and even making telephone calls. Educational institutions use the Internet for research and to deliver online courses and course material to students.

Use of the Internet has grown tremendously since its inception. The Internet's success arises from its flexibility. Instead of restricting component networks to a particular manufacturer or particular type, Internet technology allows interconnection of any kind of computer network. No network is too large or too small, too fast or too slow to be interconnected. Thus, the Internet includes inexpensive networks that can only connect a few computers within a single room as well as expensive networks that can span a continent and connect thousands of computers.

Internet service providers (ISPs) provide Internet access to customers, usually for a monthly fee. A customer who subscribes to an ISP's service uses the ISP's network to access the Internet. Because ISPs offer their services to the general public, the networks they operate are known as public access networks. In the United States, as in many countries, ISPs are private companies; in countries where telephone service is a government-regulated monopoly, the government often controls ISPs.

An organization that has many computers usually owns and operates a private network, called an intranet, which connects all the computers within the organization. To provide Internet service, the organization connects its intranet to the Internet. Unlike public access networks, intranets are restricted to provide security. Only authorized computers at the organization can connect to the intranet, and the organization restricts communication between the intranet and the global Internet. The restrictions allow computers inside the organization to exchange information but keep the information confidential and protected from outsiders.

The Internet has doubled in size every 9 to 14 months since it began in the late 1970s. In 1981 only 213 computers were connected to the Internet. By 2000 the number had grown to more than 400 million. The current number of people who use the Internet can only be estimated. Some analysts said that the number of users was expected to top 1 billion by the end of 2005.

Uses of the Internet

Before the Internet was created, the U.S. military had developed and deployed communications networks, including a network known as ARPANET. Uses of the networks were restricted to military personnel and the researchers who developed the technology. Many people regard the ARPANET as the precursor of the Internet. From the 1970s until the late 1980s the Internet was a U.S. government-funded communication and research tool restricted almost exclusively to academic and military uses. It was administered by the National Science Foundation (NSF). At universities, only a handful of researchers working on Internet research had access.

In the 1980s the NSF developed an "acceptable use policy" that relaxed restrictions and allowed faculty at universities to use the Internet for research and scholarly activities. However, the NSF policy prohibited all commercial uses of the Internet. Under this policy advertising did not appear on the Internet, and people could not charge for access to Internet content or sell products or services on the Internet.

By 1995, however, the NSF ceased its administration of the Internet. The Internet was privatized, and commercial use was permitted. This move coincided with the growth in popularity of the World Wide Web (WWW), which was developed by British physicist and computer scientist Timothy Berners-Lee. The Web replaced file transfer as the application used for most Internet traffic.

The difference between the Internet and the Web is similar to the distinction between a highway system and a package delivery service that uses the highways to move cargo from one city to another: The Internet is the highway system over which Web traffic and traffic from other applications move. The Web consists of programs running on many computers that allow a user to find and display multimedia documents (documents that contain a combination of text, photographs, graphics, audio, and video). Many analysts attribute the explosion in use and popularity of the Internet to the visual nature of Web documents. By the end of 2000, Web traffic dominated the Internet—more than 80 percent of all traffic on the Internet came from the Web.

Companies, individuals, and institutions use the Internet in many ways. Companies use the Internet for electronic commerce, also called e-commerce, including advertising, selling, buying, distributing products, and providing customer service. In addition, companies use the Internet for business-to-business transactions, such as exchanging financial information and accessing complex databases. Businesses and institutions use the Internet for voice and video conferencing and other forms of communication that enable people to *telecommute* (work away from the office using a computer).

The use of e-mail speeds communication between companies, among coworkers, and among other individuals. Media and entertainment companies run online news and weather services over the Internet, distribute music and movies, and actually broadcast audio and video, including live radio and television programs. File sharing services let individuals swap music, movies, photos, and applications, provided they do not violate copyright protections. Instant messaging enables people to exchange text messages; share digital photo, video, and audio files; and play games in real time.

HOW THE INTERNET WORKS

Internet Access

The term *Internet access* refers to the communication between a residence or a business and an ISP that connects to the Internet. Access falls into three broad categories: dedicated, dial-up, and wireless. With dedicated access, a subscriber's computer remains directly connected to the Internet at all times through a permanent, physical connection. Most large businesses have high-capacity dedicated connections; small businesses or individuals that desire dedicated access choose technologies such as digital subscriber line (DSL) or cable modems, which both use existing wiring to lower cost.

A DSL sends data across the same wires that telephone service uses, and cable modems use the same wiring that cable television uses. In each case, the electronic devices that are used to send data over the wires employ separate frequencies or channels that do not interfere with other signals on the wires. Thus, a DSL Internet connection can send data over a pair of wires at the same time the wires are being used for a telephone call, and cable modems can send data over a cable at the same time the cable is being used to receive television signals.

Another, less-popular option is satellite Internet access, in which a computer grabs an Internet signal from orbiting satellites via an outdoor satellite dish. The user usually pays a fixed monthly fee for a dedicated connection. In exchange, the company providing the connection agrees to relay data between the user's computer and the Internet.

Dial-up is the least expensive access technology, but it is

also the least convenient. To use dial-up access, a subscriber must have a telephone modem, a device that connects a computer to the telephone system and is capable of converting data into sounds and sounds back into data. The user's ISP provides software that controls the modem. To access the Internet, the user opens the software application, which causes the dial-up modem to place a telephone call to the ISP. A modem at the ISP answers the call, and the two modems use audible tones to send data in both directions.

When one of the modems is given data to send, the modem converts the data from the digital values used by computers—numbers stored as a sequence of 1s and 0s—into tones. The receiving side converts the tones back into digital values. Unlike dedicated access technologies, a dial-up modem does not use separate frequencies, so the telephone line cannot be used for regular telephone calls at the same time a dial-up modem is sending data.

How Information Travels Over the Internet

All information is transmitted across the Internet in small units of data called packets. Software on the sending computer divides a large document into many packets for transmission; software on the receiving computer regroups incoming packets into the original document. Similar to a postcard, each packet has two parts: a packet header specifying the computer to which the packet should be delivered, and a packet payload containing the data being sent. The header also specifies how the data in the packet should be combined with the data in other packets by recording which piece of a document is contained in the packet.

A series of rules known as computer communication protocols specify how packet headers are formed and how packets are processed. The set of protocols used for the Internet is named TCP/IP after the two most important protocols in the set: the Transmission Control Protocol and the Internet Protocol.

TCP/IP protocols enable the Internet to automatically detect and correct transmission problems. For example, if any network or device malfunctions, protocols detect the failure and automatically find an alternative path for packets in order to avoid the malfunction. Protocol software also ensures that data arrives complete and intact. If any packets are missing or damaged, protocol software on the receiving computer requests that the source resend them. Only when the data has arrived correctly does the protocol software make it available to the receiving application programme, and therefore to the user.

Hardware devices that connect networks in the Internet are called IP routers because they follow the IP protocol when forwarding packets. A router examines the header in each packet that arrives to determine the packet's destination. The router either delivers the packet to the destination computer across a local network or forwards the packet to another router that is closer to the final destination. Thus, a packet travels from router to router as it passes through the Internet. In some cases, a router can deliver packets across a local area wireless network, allowing desktop and laptop computers to access the Internet without the use of cables or wires. Today's business and home wireless local area networks (LANs), which operate according to a family of wireless protocols known as Wi-Fi, are fast enough to deliver Internet feeds as quickly as wired LANs.

Increasingly, cell phone and handheld computer users are also accessing the Internet through wireless cellular telephone networks. Such wide area wireless access is much slower than high-capacity dedicated, or broadband, access, or dial-up access. Also, handheld devices, equipped with much smaller screens and displays, are more difficult to use than full-sized computers. But with wide area wireless, users can access the Internet on the go and in places where access is otherwise impossible. Telephone companies are currently developing so-called 3G—for "third generation"—cellular networks that will provide wide area Internet access at DSL-like speeds.

Network Names and Addresses

To be connected to the Internet, a computer must be

assigned a unique number, known as its IP (Internet Protocol) address. Each packet sent over the Internet contains the IP address of the computer to which it is being sent. Intermediate routers use the address to determine how to forward the packet. Users almost never need to enter or view IP addresses directly. Instead, to make it easier for users, each computer is also assigned a domain name; protocol software automatically translates domain names into IP addresses.

Users encounter domain names when they use applications such as the World Wide Web. Each page of information on the Web is assigned a URL (Uniform Resource Locator) that includes the domain name of the computer on which the page is located. Other items in the URL give further details about the page. For example, the string *http* specifies that a browser should use the http protocol, one of many TCP/IP protocols, to fetch the item.

Client/Server Architecture

Internet applications, such as the Web, are based on the concept of client/server architecture. In a client/server architecture, some application programs act as information providers (servers), while other application programs act as information receivers (clients). The client/server architecture is not one-to-one. That is, a single client can access many different servers, and a single server can be accessed by a number of different clients. Usually, a user runs a client application, such as a Web browser, that contacts one server at a time to obtain information. Because it only needs to access one server at a time, client software can run on almost any computer, including small handheld devices such as personal organizers and cellular telephones. To supply information to others, a computer must run a server application.

Although server software can run on any computer, most companies choose large, powerful computers to run server software because the company expects many clients to be in contact with its server at any given time. A faster computer enables the server programme to return information with less delay.

Electronic Mail

Electronic mail, or e-mail, is a widely used Internet application that enables individuals or groups of individuals to quickly exchange messages, even if they are separated by long distances. A user creates an e-mail message and specifies a recipient using an e-mail address, which is a string consisting of the recipient's login name followed by an @ (at) sign and then a domain name. E-mail software transfers the message across the Internet to the recipient's computer, where it is placed in the specified mailbox, a file on the hard drive. The recipient uses an e-mail application to view and reply to the message, as well as to save or delete it.

Because e-mail is a convenient and inexpensive form of communication, it has dramatically improved personal and business communications. In its original form, e-mail could only be sent to recipients named by the sender, and only text messages could be sent. E-mail has been extended in two ways, and is now a much more powerful tool. Software has been invented that can automatically propagate to multiple recipients a message sent to a single address. Known as a mail gateway or list server, such software allows individuals to join or leave a mail list at any time. E-mail software has also been extended to allow the transfer of nontext documents, such as photographs and other images, executable computer programs, and prerecorded audio. Such documents, appended to an e-mail message, are called attachments.

The standard used for encoding attachments is known as Multipurpose Internet Mail Extensions (MIME). Because the Internet e-mail system only transfers printable text, MIME software encodes each document using printable letters and digits before sending it and then decodes the item when e-mail arrives. Most significantly, MIME allows a single message to contain multiple items, enabling a sender to include a cover letter that explains each of the attachments.

Other Internet Applications

Although the World Wide Web is the most popular application, some older Internet applications are still used.

For example, the Telnet application enables a user to interactively access a remote computer. Telnet gives the appearance that the user's keyboard and monitor are connected directly to the remote computer. For example, a businessperson who is visiting a location that has Internet access can use Telnet to contact their office computer. Doing so is faster and less expensive than using a dial-up modem.

Another application, known as the File Transfer Protocol (FTP), is used to download files from an Internet site to a user's computer. The FTP application is often automatically invoked when a user downloads an updated version of a piece of software.

Applications such as FTP have been integrated with the World Wide Web, making them transparent so that they run automatically without requiring users to open them. When a Web browser encounters a URL that begins with ftp://it automatically uses FTP to access the item.

Network News discussion groups (newsgroups), originally part of the Usenet network, are another form of online discussion. Thousands of newsgroups exist, on an extremely wide range of subjects. Messages to a newsgroup are not sent directly to each user. Instead, an ordered list is disseminated to computers around the world that run news server software. Newsgroup application software allows a user to obtain a copy of selected articles from a local news server or to use e-mail to post a new message to the newsgroup.

A service known as Voice Over IP (VoIP) allows individuals and businesses to make phone calls over the Internet. Low-cost services (some of them free) often transfer calls via personal computers (PCs) equipped with microphones and speakers instead of the traditional telephone handset.

But a growing number of services operate outside the PC, making calls via a special adapter that connects to a traditional telephone handset. The calls still travel over the Internet, but the person using the special adapter never has to turn on his or her computer. Thousands now use such VoIP services in lieu of traditional phone service. VoIP services are not subject to the same government regulation as traditional phone service. Thus, they are often less expensive.

BANDWIDTH

Computers store all information as binary numbers. The binary number system uses two binary digits, 0 and 1, which are called bits. The amount of data that a computer network can transfer in a certain amount of time is called the bandwidth of the network and is measured in kilobits per second (kbps) or megabits per second (mbps). A kilobit is 1 thousand bits; a megabit is 1 million bits. A dial-up telephone modem can transfer data at rates up to 56 kbps; DSL and cable modem connections are much faster and can transfer at a few mbps. The Internet connections used by businesses can operate at 45 mbps or more, and connections between routers in the heart of the Internet may operate at rates from 2,488 to 9,953 mbps (9.953 gigabits per second).

The terms *wideband* or *broadband* are used to characterize networks with high capacity, such as DSL and cable, and to distinguish them from narrowband networks, such as dial-up modems, which have low capacity. Research on dividing information into packets and switching them from computer to computer began in the 1960s. The U.S. Department of Defence Advanced Research Projects Agency (ARPA) funded a research project that created a packet switching network known as the ARPANET. ARPA also funded research projects that produced two satellite networks. In the 1970s ARPA was faced with a dilemma: Each of its networks had advantages for some situations, but each network was incompatible with the others.

ARPA focused research on ways that networks could be interconnected, and the Internet was envisioned and created to be an interconnection of networks that use TCP/IP protocols. In the early 1980s a group of academic computer scientists formed the Computer Science NETwork, which used TCP/IP protocols. Other government agencies extended the role of TCP/IP by applying it to their networks: The Department of Energy's Magnetic Fusion Energy Network

(MFENet), the High Energy Physics NETwork (HEPNET), and the National Science Foundation NETwork (NSFNET).

In the 1980s, as large commercial companies began to use TCP/IP to build private internets, ARPA investigated transmission of multimedia—audio, video, and graphics—across the Internet. Other groups investigated hypertext and created tools such as Gopher that allowed users to browse menus, which are lists of possible options. In 1989 many of these technologies were combined to create the World Wide Web.

Initially designed to aid communication among physicists who worked in widely separated locations, the Web became immensely popular and eventually replaced other tools. Also during the late 1980s, the U.S. government began to lift restrictions on who could use the Internet, and commercialization of the Internet began. In the early 1990s, with users no longer restricted to the scientific or military communities, the Internet quickly expanded to include universities, companies of all sizes, libraries, public and private schools, local and state governments, individuals, and families.

THE FUTURE OF THE INTERNET

Several technical challenges must be overcome if the Internet is to continue growing at the current phenomenal rate. The primary challenge is to create enough capacity to accommodate increases in traffic. Internet traffic is increasing as more people become Internet users and existing users send greater amounts of data. If the volume of traffic increases faster than the capacity of the network increases, congestion will occur, similar to the congestion that occurs when too many cars attempt to use a highway.

To avoid congestion, researchers have developed technologies, such as Dense Wave Division Multiplexing (DWDM), that transfer more bits per second across an optical fibre. The speed of routers and other packet-handling equipment must also increase to accommodate growth. In the short term, researchers are developing faster electronic processors; in the long term, new technologies will be required.

Another challenge involves IP addresses. Although the original protocol design provided addresses for up to 4.29 billion individual computers, the addresses have begun to run out because they were assigned in blocks. Researchers developed technologies, such as Network Address Translation (NAT), to conserve addresses. NAT allows multiple computers at a residence to "share" a single Internet address.

Other important questions concerning Internet growth relate to government controls, especially taxation and censorship. Because the Internet has grown so rapidly, governments have had little time to pass laws that control its deployment and use, impose taxes on Internet commerce, or otherwise regulate content. Many Internet users in the United States view censorship laws as an infringement on their constitutional right to free speech.

In 1996 the Congress of the United States passed the Communications Decency Act, which made it a crime to transmit indecent material over the Internet. The act resulted in an immediate outcry from users, industry experts, and civil liberties groups opposed to such censorship. In 1997 the Supreme Court of the United States declared the act unconstitutional because it violated First Amendment rights to free speech. The U.S. Congress responded in 1998 by passing a narrower antipornography bill, the Child Online Protection Act (COPA).

COPA required commercial Web sites to ensure that children could not access material deemed harmful to minors. In 1999 a federal judge blocked COPA as well, ruling that it would dangerously restrict constitutionally protected free speech. The judge's ruling was upheld by a federal appeals court on the grounds that the law's use of "community standards" in deciding what was pornographic was overly broad.

The issue reached the Supreme Court of the United States in 2002, and in a limited ruling the Supreme Court found that the community standard provision was not inherently unconstitutional. Supporters of the law welcomed the Court's ruling. However, opponents noted that the Court had sent the case back to the federal appeals court for a more comprehensive review and had ruled that the law could not go into effect until that review occurred. Some analysts who studied the various opinions written by the justices concluded that a majority of the Court was likely to find the law unconstitutional.

Increasing commercial use of the Internet has heightened security and privacy concerns. With a credit or debit card, an Internet user can order almost anything from an Internet site and have it delivered to their home or office. Companies doing business over the Internet need sophisticated security measures to protect credit card, bank account, and social security numbers from unauthorized access as they pass across the Internet. Any organization that connects its intranet to the global Internet must carefully control the access point to ensure that outsiders cannot disrupt the organization's internal networks or gain unauthorized access to the organization's computer systems and data.

Disruptions that could cause loss of life or that could be part of a coordinated terrorist attack have also become an increasing concern. For example, using the Internet to attack computer systems that control electric power grids, pipelines, water systems, or chemical refineries could cause the systems to fail, and the resulting failures could lead to fatalities and harm to the economy.

To safeguard against such attacks, the U.S. Congress passed the Homeland Security Act in November 2002. The new law creates criminal penalties, including life imprisonment, for disruptions of computer systems and networks that cause or attempt to cause death. The law also allows ISPs to reveal subscriber information to government officials without a court-approved warrant if there is a risk of death or injury.

It also enables government officials to trace e-mails and other Internet traffic during an Internet disruption without obtaining court approval. Civil liberties groups objected to the lack of court supervision of many provisions in the new law.

Chapter 4

DNA Structure

DNA stands for deoxyribonucleic acid. DNA is pretty unusual in that it is about the only common molecule capable of directing its own synthesis. The processes of mitosis and meiosis were discovered in the 1870s and 1890s. It was observed that, as cells divided, chromosomes moved around in a cell, and people began to wonder what their function was. It was determined that chromosomes were made of protein and DNA, about which people knew almost nothing.

People began to suspect that chromosomes had something to do with genetics, but couldn't explain what/how. When enough evidence was accumulated to confirm that chromosomes did, indeed, have something to do with genetics, most people thought that in some way the protein in the chromosomes served as the genetic material. People knew that DNA was also in the chromosomes, but because its structure was unknown and people didn't know much about it, few people thought it was the genetic material.

In 1928, Frederick Griffith performed an experiment using pneumonia bacteria and mice. This was one of the first experiments that hinted that DNA was the genetic code material. He used two strains of *Streptococcus pneumoniae*: a "smooth" strain which has a polysaccharide coating around it that makes it look smooth when viewed with a microscope, and a "rough" strain which doesn't have the coating, thus looks rough under the microscope.

When he injected live S strain into mice, the mice contracted pneumonia and died. When he injected live R strain, a strain which typically does not cause illness, into mice, as predicted they did not get sick, but lived. Thinking that perhaps the polysaccharide coating on the bacteria somehow caused the illness and knowing that polysaccharides are not affected by heat, Griffith then used heat to kill some of the S strain bacteria and injected those dead bacteria into mice.

This failed to infect/kill the mice, indicating that the polysaccharide coating was not what caused the disease, but rather, something within the living cell. Since Griffith had used heat to kill the bacteria and heat denatures protein, he next hypothesized that perhaps some protein within the living cells, that was denatured by the heat, caused the disease. He then injected another group of mice with a mixture of heat-killed S and live R, and the mice died! When he did a necropsy on the dead mice, he isolated live S strain bacteria from the corpses.

Griffith concluded that the live R strain bacteria must have absorbed genetic material from the dead S strain bacteria, and since heat denatures protein, the protein in the bacterial chromosomes was not the genetic material. This evidence pointed to DNA as being the genetic material. Transformation is the process whereby one strain of a bacterium absorbs genetic material from another strain of bacteria and "turns into" the type of bacterium whose genetic material it absorbed. Because DNA was so poorly understood, scientists remained skeptical up through the 1940s.

In 1952, Alfred Hershey and Martha Chase did an experiment which is so significant, it has been nicknamed the "Hershey-Chase Experiment". At that time, people knew that viruses were composed of DNA (or RNA) inside a protein coat/shell called a capsid. It was also known that viruses replicate by taking over the host cell's metabolic functions to make more virus. We are used to thinking and talking about viruses which invade our bodies and make us sick, but there are other, different kinds of viruses that infect other kinds of animals, still other viruses which infect plants, and even some viruses that infect bacteria.

A virus which infects a bacterium is called a bacteriophage because the host bacterium cell is killed as the new virus particles leave the bacterial cell. In order to do all this, the virus must inject whatever is the viral genetic code into the host cell. Thus, people realized that the viral genetic code material had to be either its DNA or its protein capsid. Hershey and Chase sought an answer to the question, "Is it the viral DNA or viral protein coat (capsid) that is the viral genetic code material which gets injected into a host bacterium cell?

To try to answer this question, Hershey and Chase performed an experiment using a bacterium named *Escherichia coli*, or *E. coli* for short (named after a scientist whose last name was Escher) and a virus called T2 that is a bacteriophage that infects *E. coli*. Isolated T2, like other viruses, is just a crystal of DNA and protein, so it must live inside *E. coli* in order to make more virus like itself. When the new T2 viruses are ready to leave the host *E. coli* cell (and go infect others), they burst the *E. coli* cell open, killing it (hence the name "bacteriophage"). The results that Hershey and Chase obtained indicated that the viral DNA, not the protein, is its genetic code material.

Hershey and Chase used radioactive chemicals to distinguish between ("label") the protein capsid and the DNA in T2 virus so they could tell which of those molecules entered the *E. coli* cells. Since some amino acids contain sulfur in their side chains, if T2 is grown in *E. coli* with a source of radioactive sulfur, the sulfur will be incorporated into the T2 protein coat making it radioactive. Since DNA has lots of phosphorus in its phosphate ($-PO_4$) groups, if T2 is grown in *E. coli* with a source of radioactive phosphorus, the phosphorus will be incorporated into the viral DNA, making that radioactive. Hershey and Chase grew two batches of T2 and *E. coli*: one with radioactive sulfur and one with radioactive phosphorus to get batches of T2 "labeled" with either radioactive S or radioactive P. Then, these radioactive T2 were placed in separate, new batches of *E. coli*, but were left there only 10 minutes.

This was to give the T2 time to inject their genetic material into the bacteria, but not reproduce. In the next step, still in separate batches, the mixtures were agitated in a kitchen blender to knock loose any viral parts not inside the *E. coli* but perhaps stuck on the outer surface. Hopefully, this would differentiate between the protein and DNA portions of the virus. Then, each mixture was spun in a centrifuge to separate the heavy bacteria (with any viral parts that had gone into them) from the liquid solution they were in (including any viral parts that had not entered the bacteria). The centrifuge causes the heavier bacteria to be pulled to the bottom of the tube where they form a pellet, while the light-weight viral "left-overs" stay suspended in the liquid portion called the supernatant.

In the subsequent step, the pellet and supernatant from each tube were separated and tested for the presence of radioactivity. Radioactive sufur was found in the supernatant, indicating that the viral protein did not go into the bacteria. Radioactive phosphorus was found in the bacterial pellet, indicating that viral DNA did go into the bacteria.

Based on these results, Hershey and Chase concluded that DNA must be the genetic code material, not protein as many poeple believed. When their experiment was published and people finally acknowledged that DNA was the genetic material, there was a lot of competition to be the first to discover its chemical structure. What was known is that DNA contains a nitrogenous base. There are two kinds of these, which include:

Pyrimidine (6-member ring of C & N)	*Purine (that + 5 member ring of C & N)*
Cytosine	Guanine
Thymine in DNAuracil in RNA 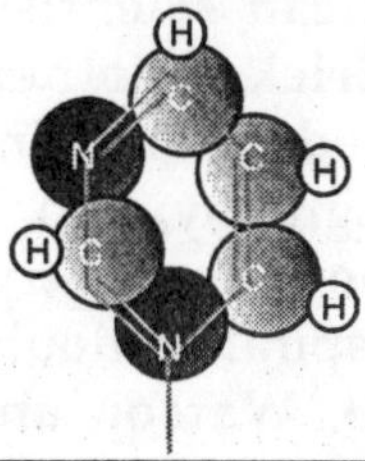	Adenine 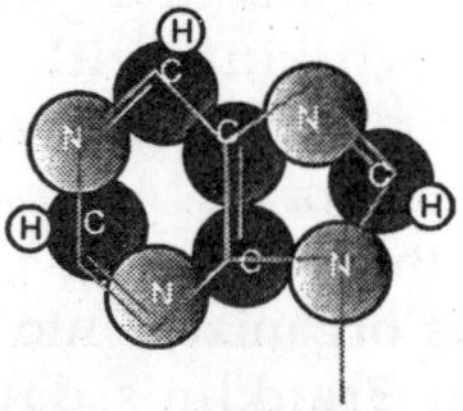

Each nitrogenous base is connected to a molecule of ribose sugar (–1 oxygen in DNA) to form a nucleoside like the *adenosine* in ATP. Each nucleoside is joined to a PO_4 (phosphate group) to form a nucleotide like adenosine monophosphate (which can be turned into ATP by adding phosphate groups). People also knew that nucleotides were somehow linked by dehydration synthesis to form DNA, but the exact structure/ arrangement was unknown.

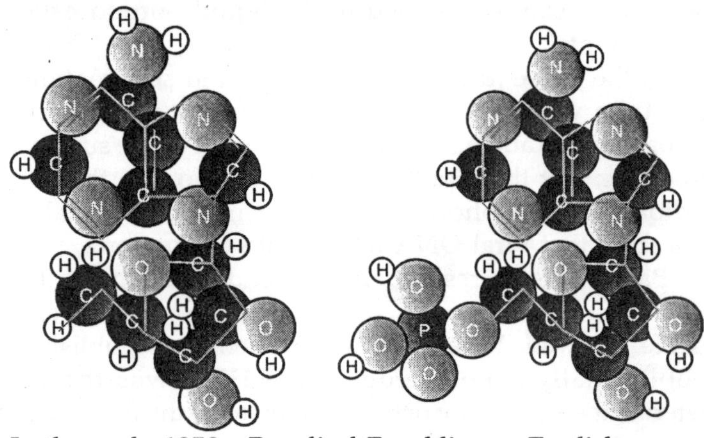

In the early 1950s, Rosalind Franklin, an Englishwoman, was doing research which involved bouncing x-rays off crystals of various substances (a process which is called x-ray crystallography), including DNA, then exposing photographic film to the x-rays. She was studying the scatter patterns made by the x-rays bouncing off the crystals of various substances (Unfortunately, she died of cancer soon afterwards, or she might have been more famous). Other people like Linus Pauling were also attempting to figure out the structure of DNA.

James Watson, a young American scientist was in England working with Francis Crick, another young researcher. Someone else showed them Franklin's photographs of DNA x-ray crystallography, and from her pictures, they were able to determine that the structure of DNA was organized into a double spiral or double helix. Based on Franklin's data, in 1953, Watson and Crick published a paper in which they proposed and described an hypothetical structure for DNA.

Subsequent research by many other people has since upheld their hypothesis, and based on subsequent examination of Franklin's lab notes and calculations, she was probably within a couple days of coming to the same conclusion when their paper was published.

For their discovery, Watson and Crick received the Nobel prize in 1962. In the intervening time, Rosalind Franklin had died in 1958 of ovarian cancer, probably due in large part to her work with x-rays. Since the Nobel prize is not awarded posthumously, people have often wondered if the Nobel committee would have included Franklin if she had still been alive.

DNA is a double helix. The outer edges are formed of alternating ribose sugar molecules and phosphate groups. The two strands go in opposite directions (1 "up" and 1 "down"). The nitrogenous bases are "inside" like rungs on a ladder. Adenine on one side pairs with thymine (uracil in RNA) on the other by hydrogen bonding, and cytosine pairs with guanine.

Note that the C-G pair has three hydrogen bonds while the A-T pair has only two, which keeps them from pairing wrong. This dictates side-to-side pairing, but says nothing about the order *along* the molecule. Watson and Crick said this variability along the molecule can account for the variety in the genetic code. Their model also accounts for how DNA can replicate itself. They said the molecule "unzips" and new matching bases are added in to create two new molecules. They called this semiconservative replication because each new molecule has one "old" and one "new" strand of DNA.

DNA codes for protein synthesis by first coding for RNA. First, the DNA code is transcribed to RNA code, which is still in the "language" of nitrogenous bases, except that adenine on the DNA pairs with uracil (in place of thymine) on the RNA. The RNA code is then translated to protein code, which is a different "language." This process involves ribosomes and two kinds of RNA: mRNA and tRNA. The mRNA codes for the gene in question and is copied off the DNA, while tRNA matches a specific group of nucleotides with a specific amino acid. A "unit" of three nucleotides on the tRNA codes for one amino acid. Each of these "units" is called an anticodon. These match up with corresponding three-nucleotide sequences on the mRNA called codons, and in this manner the amino acids are organized into the correct sequence to build a protein. The ribosome works with the mRNA and tRNA to hook the amino acids together to form a protein.

Here is a list of the mRNA codons and the corresponding amino acids for which they code.

		Second Base					
		U	C	A	G		
F	U	UUU Phe	UCU Ser	UAU Tyr	UGU Cys	U	T
i		UUC Phe	UCC Ser	UAC Try	UGC Cys	C	h
r		UUA Leu	UCA Ser	UAA Stop	UGA Stop	A	i
s		UUG Leu	UCG Ser	UAG Stop	UGG Trp	G	r
t	C	CUU Leu	CCU Pro	CAU His	CGU Arg	U	d
		CUC Leu	CCC Pro	CAC His	CGC Arg	C	
B		CUA Leu	CCA Pro	CAA Gln	CGA Arg	A	B
a		CUG Leu	CCG Pro	CAG Gln	CGG Arg	G	a
s	A	AUU Ile	ACU Thr	AAU Asn	AGU Ser	U	s
e		AUC Ile	ACC Thr	AAC Asn	AGC Ser	C	e
		AUA Ile	ACA Thr	AAA Lys	AGA Arg	A	
		AUG Met or Start	ACG Thr	AAG Lys	AGG Arg	G	
	G	GUU Val	GCU Ala	GAU Asp	GGU Gly	U	
		GUC Val	GCC Ala	GAC Asp	GGC Gly	C	
		GUA Val	GCA Ala	GAA Glu	GGA Gly	A	
		GUG Val	GCG Ala	GAG Glu	GGG Gly	G	

Mutations can be caused by a change in the sequence of the nucleotides. Some mutations have more effect than others, depending on where in the code they are and how important that area is to the code. While mutations in some areas of some genes have little effect, sickle cell anemia is caused by a mutation in only one nucleotide. This changes the codon at that location to code for a different amino acid, and that, in turn, significantly changes the shape of the hemoglobin molecules in that person's blood.

When some viruses (especially *Herpes* viruses, including Chicken Pox and Cold Sores) infect us, they insert their DNA into our cells' DNA, and stay resident in our cells for the rest of our lives. These can potentially become active again either making a person sick again (like Shingles in a person who has had Chicken Pox) or just being shed from a person's body (to infect others) without obvious symptoms of illness (like Mononucleosis). Some kinds of cancer may be caused this way. For example, there is some pretty strong evidence linking genital warts (human papillomavirus, HPV) and cervical cancer.

The AIDS virus does things "backwards." This virus contains RNA rather than DNA, yet when it gets into someone's cells, it can do reverse transcription and code from its RNA to make DNA which, then, can code to make more virus.

GENETIC ENGINEERING

We now have the knowledge and ability to transfer genes from one organism to another, which seems to have some benefits associated with it, but may also have many yet-to-be-discovered problems associated with it. Because this is all so new, not enough time has elapsed to allow scientists to study/look for any possible long-term effects of genetically-modified organisms (GMOs).

Many medicines are now made by GMOs. For example, insulin was formerly extracted from the pancreas of animals after they were slaughtered for meat. However, now most insulin is produced by bacteria with the insulin gene spliced onto their chromosome. Using this method of production, drug companies can make more insulin, faster. In theory, insulin produced in this way should be more "pure" — someone who could not use pig insulin due to a pork allergy may be able to tolerate insulin made in this way (but someone could, potentially, be allergic to some component of the bacteria present in the refined insulin). In the relatively short time that this form of insulin has been available, there is no doubt that it has saved many people's lives — let's hope that some unforseen, long-term, deleterious effects are not discovered later on.

Experimentation is being done to investigate the possible use of genetically-engineered viruses to treat genetic diseases such as cystic fibrosis. In this "treatment," a kind of virus that infects our lungs is used. The genes that enable it to infect our cells are kept while the genes which make us sick are (hopefully) all removed, and the missing human (normal) gene that relieves cystic fibrosis is then inserted into the virus' genome.

These viruses are then sprayed into the lungs of a person with CF and allowed to "infect" the cells in that person's lungs. When the normal gene is inserted into the genetic make-up of the cells lining that person's lungs, those cells function normally and the CF symptoms are alleviated. However, this treatment doesn't last because only that layer of cells is "infected," and when those cells die and are replaced by new cells, the new cells do not contain the genetic code to overcome the CF gene, and the person must inhale more genetically-engineered virus. While this seems to be a promising, life-giving, technique, no data are yet available on long-term effects and safety.

BACILLUS THURINGIENSIS

Bacillus thuringiensis (BT) is a species of bacterium that infects and kills a number of species of caterpillars. There is a species of moth whose caterpillar is a "pest" in corn plants, and insects of any kind are more successful when we humans plant huge monocultures of their favourite foods. For a number of years, now, people have realized that in certain situations, by judiciously applying BT, "pest" species of caterpillars can be infected and killed without the use of man-made, chemical insecticides.

More recently, a major US chemical company came up with the idea of creating genetically-engineered corn containing BT genes, which was good news to agri-business firms who plant huge areas of land with monocultures of corn. Because corn is wind-pollinated and therefore makes lots of pollen which, in this case, contains BT genes, many scientists are concerned about the effects of this corn on local butterfly populations.

Some research has indicated that when this pollen settles

on nearby caterpillar host plants and is, therefore, consumed by caterpillars, this might cause an increase in mortality (therefore fewer "good" butterflies such as Monarchs). It is assumed that these BT genes in corn should, they think, have no adverse effects on people or cattle who consume this corn, but no long-term testing has been done. Also, this company has convinced the government that this corn should be marketed without any labeling indicating that it is genetically-engineered — they're scared that if we are given a choice, we won't buy/eat their corn if we know it is genetically engineered.

Additionally, to increase their profits, this company has also put "suicide" genes into this corn so that farmers cannot save seed from one year to plant the next year, and have to buy more seed from them, instead. For big agri-businesses, this is of small consequence, but for small, family farmers this is total disaster. To save money, the latter often save seed from one year to plant the next, and even if they don't plant this genetically-engineered corn, if their corn is pollinated by genetically-engineered corn from a neighboring field, it will not produce viable seed.

This same chemical company came up with the idea of "transplanting" what they think is the cold-tolerance gene from a species of cold-water-inhabiting fish into tomatoes, thereby hoping to "invent" cold-tolerant tomatoes, and again, has convinced the government that these tomatoes should not be labeled in any way to indicate that they have fish genes in them, and that we should not have a choice about what we eat.

At the very least, this would be a problem for someone who is a vegetarian and chooses to not eat fish. A more serious consequence, however, would be that a person who is severely allergic to fish could also have an allergic reaction to these tomatoes and end up in the hospital (or worse...). Again, no tests were done before the government approved these tomatoes and no data are available on the long-term effects on humans (or anything else).

This same chemical company manufactures a widely-used herbicide, and came up with the idea to genetically engineer cotton, soybeans, and other crop plants so they would be immune to the effects of that herbicide. That way, farmers can (have to?) buy their seed from that company, then spray their fields with herbicide also purchased from that company to (hopefully) kill all local plants except the immune crop, simultaneously putting all their money in the corporation's pockets.

Preliminary research has shown that these resistant genes have already begun to "jump" into local weeds, thereby making them resistant to that herbicide, and again, no data are available on long-term effects on humans, other animals, or the environment in general.

DNA AND MOLECULAR GENETICS

THE PHYSICAL CARRIER OF INHERITANCE

While the period from the early 1900s to World War II has been considered the "golden age" of genetics, scientists still had not determined that DNA, and not protein, was the hereditary material. However, during this time a great many genetic discoveries were made and the link between genetics and evolution was made.

Friedrich Meischer in 1869 isolated DNA from fish sperm and the pus of open wounds. Since it came from nuclei, Meischer named this new chemical, nuclein. Subsequently the name was changed to nucleic acid and lastly to deoxyribonucleic acid (DNA). Robert Feulgen, in 1914, discovered that fuchsin dye stained DNA. DNA was then found in the nucleus of all eukaryotic cells.

During the 1920s, biochemist P.A. Levene analyzed the components of the DNA molecule. He found it contained four nitrogenous bases: cytosine, thymine, adenine, and guanine; deoxyribose sugar; and a phosphate group. He concluded that the basic unit (nucleotide) was composed of a base attached to a sugar and that the phosphate also attached to the sugar.

He (unfortunately) also erroneously concluded that the proportions of bases were equal and that there was a tetranucleotide that was the repeating structure of the molecule.

The nucleotide, however, remains as the fundemantal unit (monomer) of the nucleic acid polymer. There are four nucleotides: those with cytosine (C), those with guanine (G), those with adenine (A), and those with thymine (T).

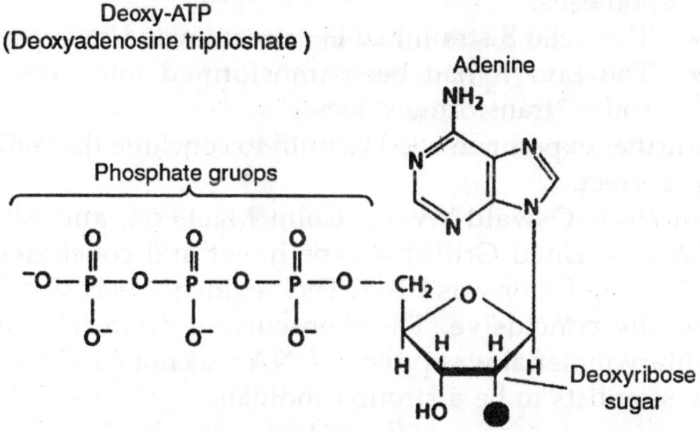

Fig. Molecular Structure of Three Nirogenous Bases

During the early 1900s, the study of genetics began in earnest: the link between Mendel's work and that of cell biologists resulted in the chromosomal theory of inheritance; Garrod proposed the link between genes and "inborn errors of metabolism"; and the question was formed: what is a gene? The answer came from the study of a deadly infectious disease: pneumonia. During the 1920s Frederick Griffith studied the difference between a disease-causing strain of the pneumonia causing bacteria (*Streptococcus peumoniae*) and a strain that did not cause pneumonia.

The pneumonia-causing strain (the S strain) was surrounded by a capsule. The other strain (the R strain) did not have a capsule and also did not cause pneumonia. Frederick Griffith was able to induce a nonpathogenic strain of the bacterium *Streptococcus pneumoniae* to become pathogenic. Griffith referred to a transforming factor that caused the non-pathogenic bacteria to become pathogenic.

Griffith injected the different strains of bacteria into mice. The S strain killed the mice; the R strain did not. He further noted that if heat killed S strain was injected into a mouse, it did not cause pneumonia. When he combined heat-killed S with Live R and injected the mixture into a mouse (remember neither alone will kill the mouse) that the mouse developed pneumonia and died. Bacteria recovered from the mouse had a capsule and killed other mice when injected into them!

Hypotheses:

- The dead S strain had been reanimated/resurrected.
- The Live R had been transformed into Live S by some "transforming factor".

Further experiments led Griffith to conclude that number 2 was correct.

In 1944, Oswald Avery, Colin MacLeod, and Maclyn McCarty revisited Griffith's experiment and concluded the transforming factor was DNA. Their evidence was strong but not totally conclusive. The then-current favourite for the hereditary material was protein; DNA was not considered by many scientists to be a strong candidate.

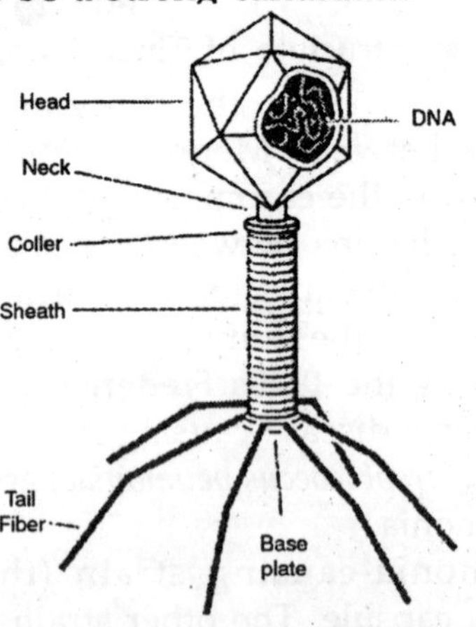

Fig. Structure of a Bacteriophage Virus

The breakthrough in the quest to determine the hereditary material came from the work of Max Delbruck and Salvador Luria in the 1940s. Bacteriophage are a type of virus that attacks bacteria, the viruses that Delbruck and Luria worked with were those attacking *Escherichia coli,* a bacterium found in human intestines. Bacteriophages consist of protein coats covering DNA. Bacteriophages infect a cell by injecting DNA into the host cell. This viral DNA then "disappears" while taking over the bacterial machinery and beginning to make new virus instead of new bacteria. After 25 minutes the host cell bursts, releasing hundreds of new bacteriophage.

Phages have DNA and protein, making them ideal to resolve the nature of the hereditary material. In 1952, Alfred D. Hershey and Martha Chase conducted a series of experiments to determine whether protein or DNA was the hereditary material. By labeling the DNA and protein with different (and mutually exclusive) radioisotopes, they would be able to determine which chemical (DNA or protein) was getting into the bacteria.

Such material must be the hereditary material (Griffith's transforming agent). Since DNA contains Phosphorous (P) but no Sulfur (S), they tagged the DNA with radioactive Phosphorous-32. Conversely, protein lacks P but does have S, thus it could be tagged with radioactive Sulfur-35. Hershey and Chase found that the radioactive S remained outside the cell while the radioactive P was found inside the cell, indicating that DNA was the physical carrier of heredity.

THE STRUCTURE OF DNA

Erwin Chargaff analyzed the nitrogenous bases in many different forms of life, concluding that the amount of purines does not always equal the amount of pyrimidines (as proposed by Levene). DNA had been proven as the genetic material by the Hershey-Chase experiments, but how DNA served as genes was not yet certain. DNA must carry information from parent cell to daughter cell. It must contain information for replicating itself. It must be chemically stable, relatively unchanging. However, it must be capable of mutational change. Without mutations there would be no process of evolution. Many scientists were interested in deciphering the structure of DNA, among them were Francis Crick, James Watson, Rosalind Franklin, and Maurice Wilkens. Watson and Crick gathered all available data in an attempt to develop a model of DNA structure. Franklin took X-ray diffraction photomicrographs of crystalline DNA extract, the key to the puzzle. The data known at the time was that DNA was a long molecule, proteins were helically coiled (as determined by the work of Linus Pauling), Chargaff's base data, and the x-ray diffraction data of Franklin and Wilkens.

DNA is a double helix, with bases to the centre (like rungs on a ladder) and sugar-phosphate units along the sides of the helix (like the sides of a twisted ladder). The strands are complementary (deduced by Watson and Crick from Chargaff's data, A pairs with T and C pairs with G, the pairs held together by hydrogen bonds). Notice that a double-ringed purine is always bonded to a single ring pyrimidine.

Purines are Adenine (A) and Guanine (G). We have encountered Adenosine triphosphate (ATP) before, although in that case the sugar was ribose, whereas in DNA it is deoxyribose. Pyrimidines are Cytosine (C) and Thymine (T). The bases are complementary, with A on one side of the molecule we only get T on the other side, similarly with G and C. If we know the base sequence of one strand we know its complement.

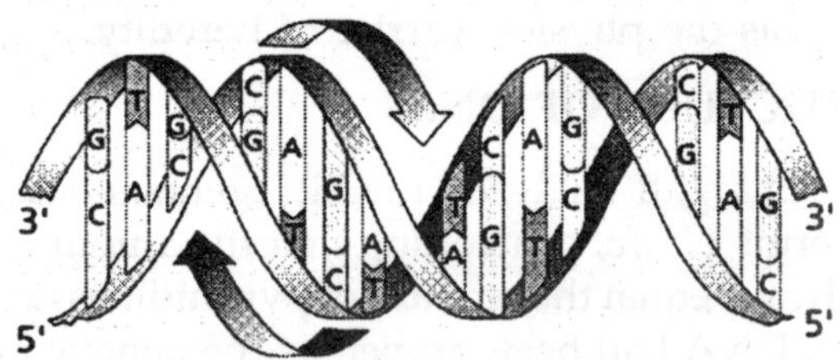

Fig. The Ribbon Model of DNA

DNA REPLICATION

DNA was proven as the hereditary material and Watson et al. had deciphered its structure. What remained was to determine how DNA copied its information and how that was expressed in the phenotype. Matthew Meselson and Franklin W. Stahl designed an experiment to determine the method of DNA replication. Three models of replication were considered likely.

- *Conservative replication:* Would somehow produce an entirely new DNA strand during replication.

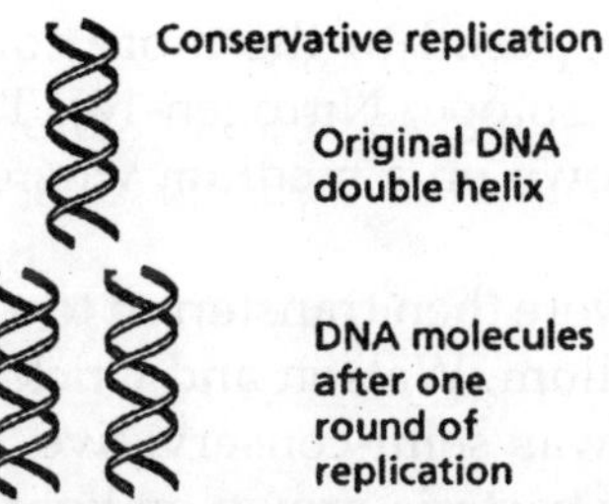

Fig. Conservative Model of DNA Replication

- *Semiconservative Replication:* Would produce two DNA molecules, each of which was composed of one-half of the parental DNA along with an entirely new complementary strand. In other words the new DNA would consist of one new and one old strand of DNA. The existing strands would serve as complementary templates for the new strand.

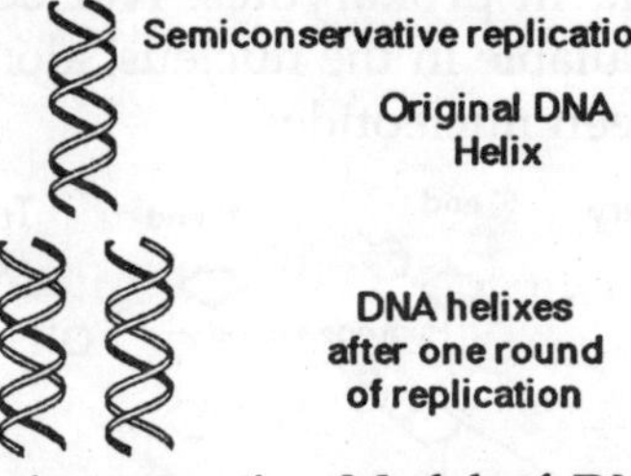

Fig. The Semiconservative Model of DNA Structure

- *Dispersive replication:* Involved the breaking of the parental strands during replication, and somehow, a reassembly of molecules that were a mix of old and new fragments on each strand of DNA.

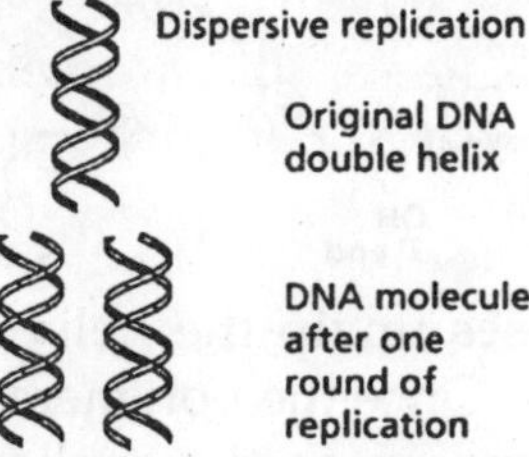

Fig. The Dispersive Replication Model of DNA Replication

The Meselson-Stahl experiment involved the growth of *E. coli* bacteria on a growth medium containing heavy nitrogen

(Nitrogen-15 as opposed to the more common, but lighter molecular weight isotope, Nitrogen-14). The first generation of bacteria was grown on a medium where the sole source of N was Nitrogen-15.

The bacteria were then transferred to a medium with light (Nitrogen-14) medium. Watson and Crick had predicted that DNA replication was semi-conservative. If it was, then the DNA produced by bacteria grown on light medium would be intermediate between heavy and light. It was. DNA replication involves a great many building blocks, enzymes and a great deal of ATP energy (remember that after the S phase of the cell cycle cells have a G phase to regenerate energy for cell division).

Only occurring in a cell once per (cell) generation, DNA replication in humans occurs at a rate of 50 nucleotides per second, 500/second in prokaryotes. Nucleotides have to be assembled and available in the nucleus, along with energy to make bonds between nucleotides.

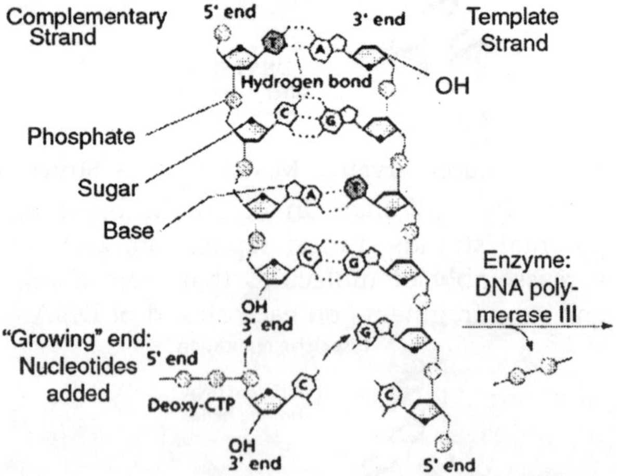

DNA polymerases unzip the helix by breaking the H-bonds between bases. Once the polymerases have opened the molecule, an area known as the replication bubble forms (always initiated at a certain set of nucleotides, the origin of replication). New nucleotides are placed in the fork and link to the corresponding parental nucleotide already there (A with

T, C with G). Prokaryotes open a single replication bubble, while eukaryotes have multiple bubbles. The entire length of the DNA molecule is replicated as the bubbles meet.

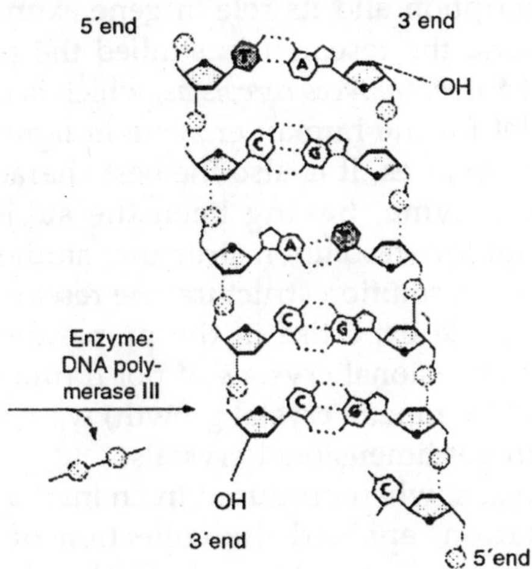

Fig. The Roles of DNA Polymerases

Since the DNA strands are antiparallel, and replication proceeds in thje 5' to 3' direction on EACH strand, one strand will form a continuous copy, while the other will form a series of short Okazaki fragments.

ENZYME STRUCTURE PROVIDES CLUES TO DNA

Before a cell can begin to divide or differentiate, the genetic information within the cell's DNA must be copied, or "transcribed," onto complementary strands of RNA. RNA polymerase II (pol II) is an enzyme that, by itself, can unwind the DNA double helix, synthesize RNA, and proofread the result. When combined with other molecules that regulate and control the transcription process, pol II is the key to successful interpretion of an organism's genetic code.

However, the size, complexity, scarcity, and fragility of pol II complexes have made analysis of these macromolecules by x-ray crystallography a formidable challenge. A team of structural biologists has met this challenge using data obtained from both the Stanford Synchrotron Radiation Laboratory and the Macromolecular Crystallography Facility at the ALS. The resultant high-resolution model of a 10-subunit pol II complex suggests roles for each of the subunits and will allow researchers to begin unraveling the intricacies of DNA transcription and its role in gene expression.

In this work, the researchers studied the pol II enzyme from the yeast *Saccharomyces cerevisiae*, which is likely to be an excellent model for the human enzyme in light of its highly similar gene sequences. It is also the best-characterized form of the pol II enzyme, having been the subject of many biochemical and low-resolution structural studies in the past. To obtain a high-resolution structure, the research team drew on its considerable expertise in the preparation of protein crystals: two-dimensional crystals of pol II (minus two small subunits found to impede crystal growth) were used as seeds for growing three-dimensional crystals.

These crystals, when produced in an inert atmosphere to prevent oxidation, enabled the collection of data to 3.5-angstrom resolution. The addition of a final soaking procedure to produce uniform crystals, combined with high-brightness x-ray sources, resulted in a resolution of 3.0 angstroms.

The current results bring into focus the somewhat fuzzy features previously observed in or inferred from earlier experiments. More importantly, the structural details suggest possible explanations for some of the unusual characteristics of this enzyme, which include a high processivity (the ability to synthesize very long strands of RNA) and the tendency to work in periodic spurts separated by pauses.

While it is known that additional proteins (transcription factors) play a role in controlling the activity of pol II (for example, restarting after a pause), scientists have yet to understand how such proteins interact with pol II binding sites to perform their various functions. The pol II model reported here establishes the positions of the various subunits and provides detailed information about the DNA/RNA binding domains.

The data reveal two main subunits (Rpb1 and Rpb2) separated by a deep cleft where DNA can enter the complex. At the end of the cleft is the active site, where the DNA can be unwound for a short distance (the "transcription bubble") and a DNA/RNA hybrid can be produced. Two prominent grooves lead away from the active site, either of which could accommodate the exiting RNA transcript. An opening below the active site may allow the entry of nucleotides (for manufacturing RNA) and transcription factors (for regulating the process). The same opening may provide room for the leading end of the RNA strand during "backtracking" maneuvers, which are important for proofreading and for traversing obstacles such as DNA damage.

Other notable features that might help account for the great stability of this transcribing complex include a pair of "jaws" that appear to grip the DNA strands as they enter the complex and, closer to the active site, a clamp on the DNA that could possibly be locked in the closed position by the presence of RNA.

The high-resolution pol II structure reported here is a landmark achievement, pulling together threads from numerous diverse research efforts into a cohesive whole. Further study should yield many new insights into the detailed mechanisms of pol II and its transcription factors. Construction of an atomic model is already well underway.

UNRAVELING DNA

Encoded into the double-helical strands of DNA, the human genome is the complete set of instructions required to make a human being. While the mapping of the human genome (roughly three billion components) is certainly a Herculean accomplishment, it is only the first step toward realizing the full potential of genomic medicine in the diagnosis, monitoring, and treatment of disease. Beyond knowing what the genetic blueprint says, scientists must understand how that blueprint gets interpreted, or "expressed" as an individual with unique traits. The pol II enzyme is the catalyst for a major step in this process.

As a pol II molecule slides along a DNA molecule, it "unzips" the strands of the DNA double helix, synthesizes a complementary strand of RNA (which will carry the genetic information to where it is needed), and verifies that no mistakes have occurred. This process is regulated by transcription factors—separate molecules that bind to pol II and determine which genes are expressed, at what stage of development, and in which tissue.

Done correctly, this process results in healthy cell growth and differentiation; otherwise, aberrations such as cancer can be the result. Thus, details of the structure of pol II, including information about its binding sites and how they interact with transcription factors, will provide valuable insight into the detailed mechanisms underlying the flow of genetic information from DNA to RNA to protein, which is necessary for life and health.

Chapter 5

RNA Structure

RNA CONTAINS RIBOSE AND URACIL

We now turn our attention to RNA, which differs from DNA in three respects. First, the backbone of RNA contains ribose rather than 2′-deoxyribose. That is, ribose has a hydroxyl group at the 2′ position. Second, RNA contains uracil in place of thymine. Uracil has the same single-ringed structure as thymine, except that it lacks the 5′ methyl group. Thymine is in effect 5′methyl-uracil. Third, RNA is usually found as a single polynucleotide chain. Except for the case of certain viruses, RNA is not the genetic material and does not need to be capable of serving as a template for its own replication. Rather, RNA functions as the intermediate, the mRNA, between the gene and the protein-synthesizing machinery.

Another function of RNA is as an adaptor, the tRNA, between the codons in the mRNA and amino acids. RNA can also play a structural role as in the case of the RNA components of the ribosome. Yet another role for RNA is as a regulatory molecule, which through sequence complementarity binds to, and interferes with the translation of, certain mRNAs. Finally, some RNAs (including one of the structural RNAs of the ribosome) are enzymes that catalyze essential reactions in the cell. In all of these cases, the RNA is copied as a single strand off only one of the two strands of the DNA template, and its complementary strand does not exist. RNA is capable of forming long double helices, but these are unusual in nature.

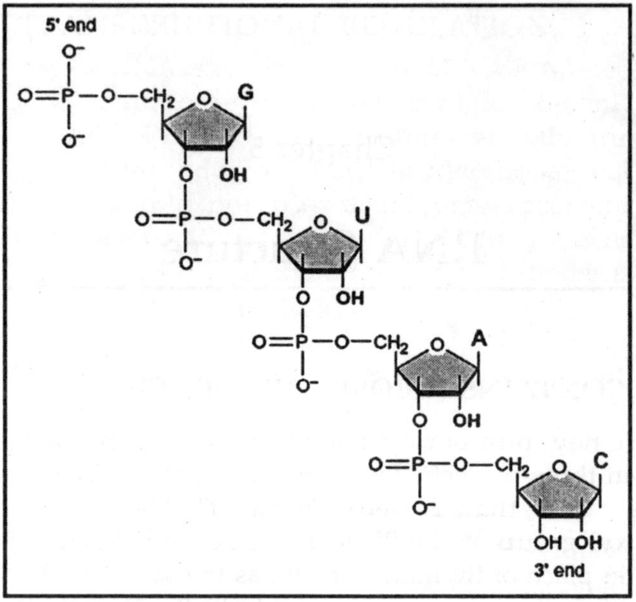

Fig. Structural Features of RNA

RNA CHAINS FOLD BACK

Despite being single-stranded, RNA molecules often exhibit a great deal of double-helical character. This is because RNA chains frequently fold back on themselves to form base-paired segments between short stretches of complementary sequences. If the two stretches of complementary sequence are near each other, the RNA may adopt one of various stem-loop structures in which the intervening RNA is looped out from the end of the double-helical segment as in a hairpin, a bulge, or a simple loop.

The stability of such stem-loop structures is in some instances enhanced by the special properties of the loop. For example, a stem-loop with the "tetraloop" sequence UUCG is unexpectedly stable due to special base-stacking interactions in the loop. Base pairing can also take place between sequences that are not contiguous to form complex structures aptly named pseudoknots.

The regions of base pairing in RNA can be a regular

double helix or they can contain discontinuities, such as noncomplementary nucleotides that bulge out from the helix. A feature of RNA that adds to its propensity to form double-helical structures is an additional, non-Watson-Crick base pair. This is the G:U base pair, which has hydrogen bonds between N_3 of uracil and the carbonyl on C_6 of guanine and between the carbonyl on C_2 of uracil and N1 of guanine. Because G:U base pairs can occur as well as the four conventional, Watson-Crick base pairs, RNA chains have an enhanced capacity for self-complementarity.

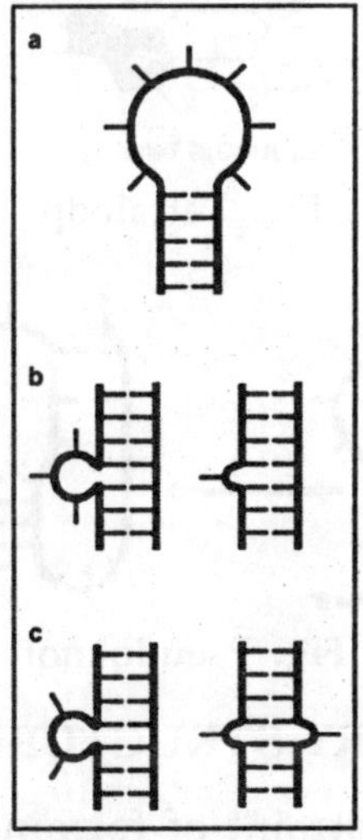

Fig. Double Helical Characteristics of RNA

Thus, RNA frequently exhibits local regions of base pairing but not the long-range, regular helicity of DNA. The presence of 2'-hydroxyls in the RNA backbone prevents RNA from adopting a B-form helix. Rather, double-helical RNA resembles the A-form structure of DNA. As such, the minor groove is wide and shallow, and hence accessible, but recall that the minor groove offers little sequence-specific information. Meanwhile, the major groove is so narrow and deep that it is not very accessible to amino acid side chains from interacting proteins.

Thus, the RNA double helix is quite distinct from the

DNA double helix in its detailed atomic structure and less well suited for sequence-specific interactions with proteins (although some proteins do bind to RNA in a sequence-specific manner).

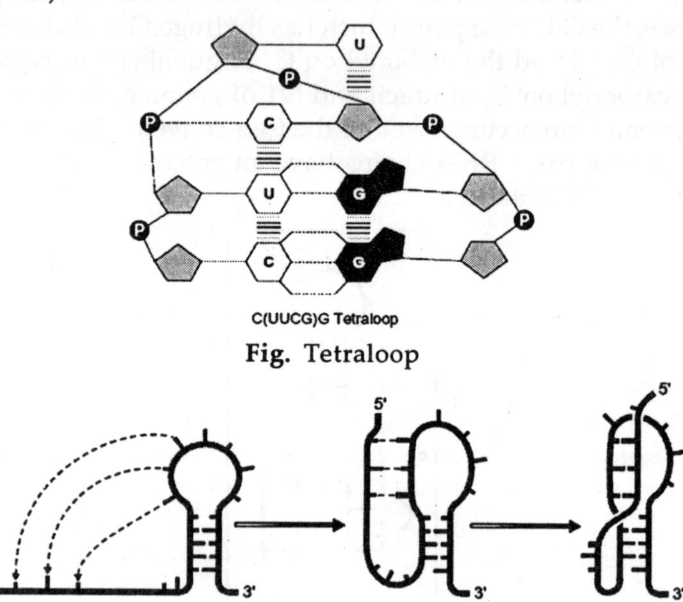

Fig. Tetraloop

Fig. Pseudoknot

COMPLEX TERTIARY STRUCTURES OF RNA

Freed of the constraint of forming long-range regular helices, RNA can adopt a wealth of tertiary structures. This is because RNA has enormous rotational freedom in the backbone of its non-base-paired regions. Thus, RNA can fold up into complex tertiary structures frequently involving unconventional base pairing, such as the base triples and base-backbone interactions seen in tRNAs.

Proteins can assist the formation of tertiary structures by large RNA molecules, such as those found in the ribosome. Proteins shield the negative charges of backbone phosphates, whose electrostatic repulsive forces would otherwise destabilize the structure.

Researchers have taken advantage of the potential

structural complexity of RNA to generate novel RNA species (not found in nature) that have specific desirable properties. By synthesizing RNA molecules with randomized sequences, it is possible to generate mixtures of oligonucleotides representing enormous sequence diversity.

For example, a mixture of oligoribonucleotides of length 20 and having four possible nucleotides at each position would have a potential complexity of 420 sequences or 1012 sequences! From mixtures of diverse oligoribonucleotides, RNA molecules can be selected biochemically that have particular properties, such as an affinity for a specific small molecule.

SOME RNAS ARE ENZYMES

It was widely believed for many years that only proteins could be enzymes. An enzyme must be able to bind a substrate, carry out a chemical reaction, release the product and repeat this sequence of events many times. Proteins are well suited to this task because they are composed of many different kinds of amino acids and they can fold into complex tertiary structures with binding pockets for the substrate and small molecule cofactors and an active site for catalysis. Now we know that RNAs, which as we have seen can similarly adopt complex tertiary structures, can also be biological catalysts.

Such RNA enzymes are known as ribozymes, and they exhibit many of the features of a classical enzyme, such as an active site, a binding site for a substrate and a binding site for a cofactor, such as a metal ion. One of the first ribozymes to be discovered was RNase P, a ribonuclease that is involved in generating tRNA molecules from larger, precursor RNAs. RNase P is composed of both RNA and protein; however, the RNA moiety alone is the catalyst. The protein moiety of RNase P facilitates the reaction by shielding the negative charges on the RNA so that it can bind effectively to its negatively charged substrate.

The RNA moiety is able to catalyze cleavage of the tRNA precursor in the absence of the protein if a small, positively charged counter ion, such as the peptide spermidine, is used to shield the repulsive, negative charges. Other ribozymes carry out trans-esterification reactions involved in the removal of intervening sequences known as introns from precursors to certain mRNAs, tRNAs, and ribosomal RNAs in a process known as RNA splicing.

THE HAMMERHEAD RIBOZYME

Before concluding our discussion of RNA, let us look in more detail at the structure and function of one particular ribozyme, the hammerhead. The hammerhead is a sequence-specific ribonuclease that is found in certain infectious RNA agents of plants known as *viroids*, which depend on self-cleavage to propagate.

When the viroid replicates, it produces multiple copies of itself in one continuous RNA chain. Single viroids arise by cleavage, and this cleavage reaction is carried out by the RNA sequence around the junction. One such self-cleaving sequence is called the *hammerhead* because of the shape of its secondary structure, which consists of three base-paired stems surrounding a core of non-complementary nucleotides required for catalysis.

The tertiary structure of the ribozyme, however, looks more like a wishbone. To understand how the hammerhead works, let us first look at how RNA undergoes hydrolysis under alkaline conditions. At high pH, the 2′ hydroxyl of the ribose in the RNA backbone can become deprotonated, and the resulting negatively charged oxygen can attack the scissile phosphate at the 3′ position of the same ribose. This reaction breaks the RNA chain, producing a 2′, 3′ cyclic phosphate and a free 5′ hydroxyl. Each ribose in an RNA chain can undergo this reaction, completely cleaving the parent molecule into nucleotides. Many protein ribonucleases also cleave their RNA substrates via the formation of a 2′, 3′ cyclic phosphate. Working at normal cellular pH, these protein enzymes use a metal ion, bound at their active site, to activate the 2′ hydroxyl of the RNA.

The hammerhead is a sequencespecific ribonuclease, but it too cleaves RNA via the formation of a 2′, 3′ cyclic phosphate. Hammerhead-mediated cleavage involves a ribozyme-bound Mg'' ion that deprotonates the 2' hydroxyl at neutral pH, resulting in nucleophilic attack on the scissile phosphate. Because the normal reaction of the hammerhead is self-cleavage, it is not really a catalyst; each molecule normally promotes a reaction one time only, thus having a turnover number of one.

But the hammerhead can be engineered to function as a true ribozyme by dividing the molecule into two portions—one, the ribozyme, that contains the catalytic core and the other, the substrate, that contains the cleavage site. The substrate binds to the ribozyme at stems I and III. After cleavage, the substrate is released and replaced by a fresh uncut substrate, thereby allowing repeated rounds of cleavage.

LIFE EVOLVE FROM AN RNA

The discovery of ribozymes has profoundly altered our view of how life might have evolved. We can now imagine that there was a primitive form of life based entirely on RNA. In this world, RNA would have functioned as the genetic material and as the enzymatic machines. This RNA world would have preceded life as we know it today, in which information transfer is based on DNA, RNA, and protein. A hint that the protein world might have arisen from an RNA world is the discovery that the component in the ribosome that is responsible for the formation of the peptide bond, the peptidyl transferase, is an RNA molecule.

Unlike RNase P, the hammerhead, and other previously known ribozymes which act on phosphorous centers, the peptidyl transferase acts on a carbon centre to create the peptide bond. It thus links RNA chemistry to the most fundamental reaction in the protein world, peptide bond formation. Perhaps then the ribosome ribozyme is a relic of an earlier form of life in which all enzymes were RNAs. DNA is usually in the form of a right-handed double helix. The helix consists of two polydeoxynucleotide chains. Each chain is an alternating polymer of deoxyribose sugars and phosphates that are joined together via phosphodiester linkages.

One of four bases protrudes from each sugar: adenine and guanine, which are purines, and thymine and cytosine, which are pyrimidines. While the sugar phosphate backbone is regular, the order of bases is irregular and this is responsible for the information content of DNA.

Each chain has a 5′ to 3′ polarity, and the two chains of the double helix are oriented in an antiparallel manner—that is, they run in opposite directions. Pairing between the bases holds the chains together. Pairing is mediated by hydrogen bonds and is specific: Adenine on one chain is always paired with thymine on the other chain, whereas guanine is always paired with cytosine.

This strict base-pairing reflects the fixed locations of hydrogen atoms in the purine and pyrimidine bases in the forms of those bases found in DNA. Adenine and cytosine almost always exist in the amino as opposed to the imino tautomeric forms, whereas guanine and thymine almost always exist in the keto as opposed to enol forms. The complementarity between the bases on the two strands gives DNA its self-coding character. The two strands of the double helix fall apart (denature) upon exposure to high temperature, extremes of pH, or any agent that causes the breakage of hydrogen bonds. Upon slow return to normal cellular conditions, the denatured single strands can specifically reassociate to biologically active double helices (renature or anneal).

DNA in solution has a helical periodicity of about 10.5 base pairs per turn of the helix. The stacking of base pairs upon each other creates a helix with two grooves. Because the sugars protrude from the bases at an angle of about 120°, the grooves are unequal in size. The edges of each base pair are exposed in the grooves, creating a pattern of hydrogen bond donors and acceptors and of van der Waals surfaces that identifies the base pair. The wider—or *major*—groove is richer in chemical information than the narrow (*minor*) groove and is more important for recognition by nucleotide sequence-specific binding proteins. Almost all cellular DNAs are extremely long molecules, with only one DNA molecule within a given chromosome. Eukaryotic cells accommodate this extreme length in part by wrapping the DNA around protein particles known as nucleosomes.

Most DNA molecules are linear but some DNAs are circles, as is often the case for the chromosomes of prokaryotes and for certain viruses. DNA is flexible. Unless the molecule is topologically constrained, it can freely rotate to accommodate changes in the number of times the two strands twist about each other. DNA is topologically constrained when it is in the form of a covalently closed circle, or when it is entrained in chromatin. The linking number is an invariant topological property of covalently closed circular DNA. It is the number of times one strand would have to be passed through the other strand in order to separate the two circular strands.

The linking number is the sum of two interconvertible geometric properties: twist, which is the number of times the two strands are wrapped around each other; and the writhing number, which is the number of times the long axis of the DNA crosses over itself in space.

DNA is relaxed under physiological conditions when it has about 10.5 base pairs per turn and is free of writhe. If the linking number is decreased, then the DNA becomes torsionally stressed, and it is said to be negatively supercoiled. DNA in cells is usually negatively supercoiled by about 6%. The left-handed wrapping of DNA around nucleosomes introduces negative supercoiling in eukaryotes. In prokaryotes, which lack histones, the enzyme DNA gyrase is responsible for generating negative supercoils. DNA gyrase is a member of the type II family of topoisomerases. These enzymes change the linking number of DNA in steps of two by making a transient break in the double helix and passing a region of duplex DNA through the break.

Some type II topoisomerases relax supercoiled DNA, whereas DNA gyrase generates negative supercoils. Type I topoisomerases also relax supercoiled DNAs but do so in steps of one in which one DNA strand is passed through a transient nick in the other strand. RNA differs from DNA in the following ways: its backbone contains ribose rather than 2'-

deoxyribose; it contains the pyrimidine uracil in place of thymine; and it usually exists as a single polynucleotide chain, without a complementary chain. As a consequence of being a single strand, RNA can fold back on itself to form short stretches of double helix between regions that are complementary to each other. RNA allows a greater range of base pairing than does DNA. Thus, as well as A:U and C:G pairing, U can also pair with G. This capacity to form a non-Watson-Crick base pair adds to the propensity of RNA to form doublehelical segments.

Freed of the constraint of forming longrange regular helices, RNA can form complex tertiary structures, which are often based on unconventional interactions between bases and between bases and the sugarphosphate backbone. Some RNAs act as enzymes—they catalyze chemical reactions in the cell and in vitro.

These RNA enzymes are known as ribozymes. Most ribozymes act on phosphorous centers, as in the case of the ribonuclease RNase P. RNase P is composed of protein and RNA, but it is the RNA moiety that is the catalyst. The hammerhead is a self-cleaving RNA, which cuts the RNA backbone via the formation of a 2′, 3′ cyclic phosphate in a reaction that involves an RNA-bound Mg″ ion. Peptidyl transferase is an example of a ribozyme that acts on a carbon centre. This ribozyme, which is responsible for the formation of the peptide bond, is one of the RNA components of the ribosome.

The discovery of RNA enzymes that can act on phosphorous or carbon centers suggests that life might have evolved from a primitive form in which RNA functioned both as the genetic material and as the enzymatic machinery.

Chapter 6

DNA Replication

DNA replication is the process of copying a double-stranded DNA molecule to form two double-stranded molecules. The process of DNA replication is a fundamental process used by all living organisms as it is the basis for biological inheritance.

As each DNA strand holds the same genetic information, both strands can serve as templates for the reproduction of the opposite strand. The template strand is preserved in its entirety and the new strand is assembled from nucleotides. This process is called "semiconservative replication". The resulting double-stranded DNA molecules are identical; proofreading and error-checking mechanisms exist to ensure near perfect fidelity. In a cell, DNA replication must happen before cell division can occur.

DNA synthesis begins at specific locations in the genome, called "origins", where the two strands of DNA are separated. RNA primers attach to single stranded DNA and the enzyme DNA polymerase extends the primers to form new strands of DNA, adding nucleotides matched to the template strand. The unwinding of DNA and synthesis of new strands forms a replication fork. In addition to DNA polymerase, a number of other proteins are associated with the fork and assist in the initiation and continuation of DNA synthesis.

DNA replication can also be performed artificially, using the same enzymes used within the cell. DNA polymerases and artificial DNA primers are used to initiate DNA synthesis at known sequences in a template molecule. The polymerase chain reaction (PCR), a common laboratory technique, employs artificial synthesis in a cyclic manner to rapidly and specifically amplify a target DNA fragment from a pool of DNA.

DNA REPLICATION IS SEMI-CONSERVATIVE

DNA replication of one helix of DNA results in two identical helices. If the original DNA helix is called the "parental" DNA, the two resulting helices can be called "daughter" helices. Each of these two daughter helices is a nearly exact copy of the parental helix (it is not 100% the same due to mutations).

DNA creates "daughters" by using the parental strands of DNA as a template or guide. Each newly synthesized strand of DNA (daughter strand) is made by the addition of a nucleotide that is complementary to the parent strand of DNA. In this way, DNA replication is semi-conservative, meaning that one parent strand is always passed on to the daughter helix of DNA.

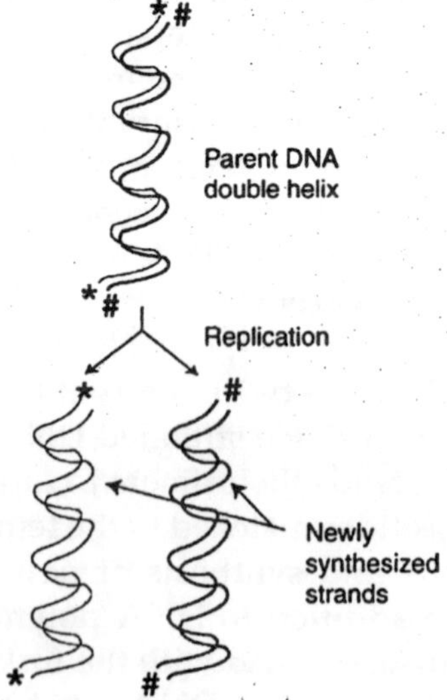

Fig. The Semi-Conservative Nature of DNA Replication

REPLICATION FORKS

The first step in DNA replication is the separation of the two DNA strands that make up the helix that is to be copied.

DNA Helicase untwists the helix at locations called replication origins. The replication origin forms a Y shape, and is called a replication fork. The replication fork moves down the DNA strand, usually from an internal location to the strand's end. The result is that every replication fork has a twin replication fork, moving in the opposite direction from that same internal location to the strand's opposite end.

Single-stranded binding proteins (SSB) work with helicase to keep the parental DNA helix unwound. It works by coating the unwound strands with rigid subunits of SSB that keep the strands from snapping back together in a helix. The SSB subunits coat the single-strands of DNA in a way as not to cover the bases, allowing the DNA to remain available for base-pairing with the newly synthesized daughter strands.

The two parent strands of DNA are separated to begin replication, one strand is oriented in the 5' to 3' direction while the other strand is oriented in the 3' to 5' direction. DNA replication, however, is inflexible: the enzyme that carries out the replication, DNA polymerase, only functions in the 5' to 3' direction.

This characteristic of DNA polymerase means that the daughter strands synthesize through different methods, one adding nucleotides one by one in the direction of the replication fork, the other able to add nucleotides only in chunks. The first strand, which replicates nucleotides one by one is called the leading strand; the other strand, which replicates in chunks, is called the lagging strand.

THE LEADING AND LAGGING STRANDS

THE LEADING STRAND

Since DNA replication moves along the parent strand in the 5' to 3' direction, replication can occur very easily on the leading strand. The nucleotides are added in the 5' to 3' direction. Triggered by RNA primase, which adds the first nucleotide to the nascent chain, the DNA polymerase simply sits near the replication fork, moving as the fork does, adding nucleotides one after the other, preserving the proper anti-parallel orientation. This sort of replication, since it involves one nucleotide being placed right after another in a series, is called continuous.

THE LAGGING STRAND

Whereas the DNA polymerase on the leading strand can simply follow the replication fork, because DNA polymerase must move in the 5' to 3' direction, on the lagging strand the enzyme must move away from the fork. But if the enzyme moves away from the fork, and the fork is uncovering new DNA that needs to be replicated, then how can the lagging strand be replicated at all?

The problem posed by this question is answered through an ingenious method. The lagging strand replicates in small segments, called Okazaki fragments. These fragments are stretches of 100 to 200 nucleotides in humans (1000 to 2000 in bacteria) that are synthesized in the 5' to 3' direction away from the replication fork.

Yet while each individual segment is replicated away from the replication fork, each subsequent Okazaki fragment is replicated more closely to the receding replication fork than the fragment before. These fragments are then stitched together by DNA ligase, creating a continuous strand. This type of replication is called discontinuous

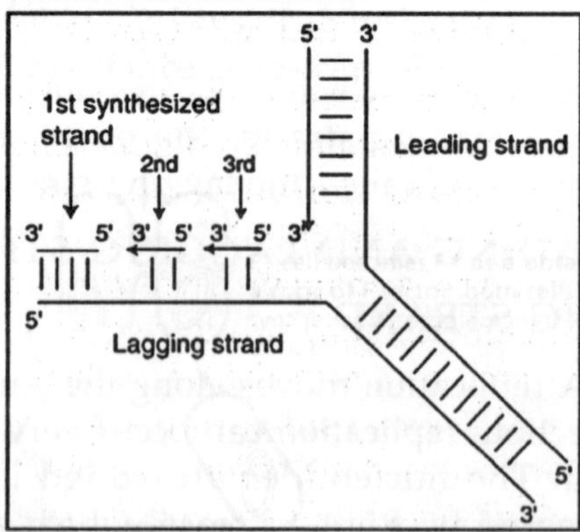

Fig. Leading and Lagging Strands

The first synthesized Okazaki fragment on the lagging strand is the furthest away from the replication fork, which is itself receding to the right. Each subsequent Okazaki fragment starts at the replication fork and continues until it meets the previous fragment. The two fragments are then stitched together by DNA ligase.

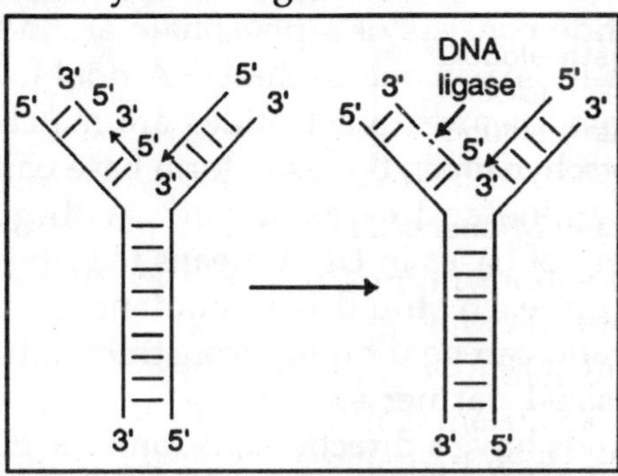

Fig. Patching Up Okazaki Fragments

We can also see how replication on the lagging strand remains slightly behind that on the leading strand. Because synthesis on the lagging strand takes place in a "backstitching" mechanism, its replication is slightly delayed in relation to synthesis on the leading strand. The lagging strand must wait for a patch of the parent helix to open up a short distance in front of the newly synthesized strand before it can begin its synthesis back to the end of the daughter strand. This "Lag" time does not occur in the leading strand because it synthesizes the new strand by following right behind as the helix unwinds at the replication fork.

Another complication to replication on the lagging strand is the initiation of replication. Whereas the RNA primer on the leading strand only has to trigger the initiation of the strand once, on the lagging strand each individual Okazaki fragment must be triggered. On the lagging strand, then, an enzyme called primase that moves with the replication fork synthesizes numerous RNA primers, each of which triggers the growth of an Okazaki fragment. The RNA primers are eventually removed leaving gaps that are filled by the replication machinery.

DNA STRUCTURE

DNA usually exists in a double-stranded structure, with

both strands coiled together to form the characteristic double-helix. Each single strand of DNA is a chain of four types of nucleotide: adenine, cytosine, guanine, and thymine.

A nucleotide consists of a phosphate and a deoxyribose sugar forming the backbone of the DNA double helix plus a base that points inwards. Nucleotides are matched between strands through hydrogen bonds to form base pairs. Adenine pairs with thymine and cytosine pairs with guanine. The physical pairing of bases in DNA means that the information contained within each strand is redundant. The nucleotides on a single strand can be used to reconstruct nucleotides on a newly synthesized partner strand.

DNA strands have a directionality, and the different ends of a single strand are called the "3' end" and the "5' end" (these refer to the carbon atom in ribose that the next phosphate in the chain attaches to). In addition to being complementary, the two strands of DNA are antiparallel: they are orientated in opposite directions. This directionality has consequences in DNA synthesis, because DNA polymerase can only synthesize DNA in one direction by adding nucleotides to the 3' end of a DNA strand.

DNA POLYMERASE

DNA polymerases are a family of enzymes critical for all forms of DNA replication. A DNA polymerase synthesizes a new strand of DNA by extending the 3' end of an existing nucleotide chain, adding new nucleotides matched to the template strand one at a time.

Some DNA polymerases may also have some proofreading ability, removing nucleotides from the end of a strand in order to remove any mismatched bases. DNA polymerases are generally extremely accurate, making less than one error for every million nucleotides added.

The energy for the process of DNA polymerization comes from the two additional phosphates attached to each of the unincorporated nucleotides. These free nucleotides, also known as nucleoside triphosphates, contain a total of three phosphates. When a nucleotide is being added to a growing DNA strand, two of the phosphates are removed and the energy produced is used to attach the remaining phosphate to the growing chain. The energetics of this process may also explain the directionality of synthesis - if DNA were synthesized in the 3' to 5' direction, the energy for the process would come from the 5' end of the growing strand rather than from free nucleotides. During proofreading, if the 5' nucleotide needed to be removed this triphosphate end would be lost, losing the energy source required to add a new nucleotide to the end.

DNA polymerase can only extend an existing DNA strand paired with a template strand, it cannot begin the synthesis of a new strand. To do this a short fragment of DNA or RNA, called a primer, must be created and paired with the template strand before DNA polymerase can synthesize new DNA.

DNA REPLICATION WITHIN THE CELL

Origins of Replication

For a cell to divide, it must first replicate its DNA. This process is initiated at particular points within the DNA, known as "origins", which are targeted by proteins that separate the two strands and initiate DNA synthesis. Origins contain DNA sequences recognized by replication initiator proteins (eg. dnaA in *E coli'* and the Origin Recognition Complex in yeast). These initiator proteins recruit other proteins to separate the two strands and initiate replication forks. Initiator proteins recruit other proteins to separate the DNA strands at the origin, forming a bubble.

Origins tend to be "AT-rich" (rich in adenine and thymine bases) to assist this process because A-T base pairs have two hydrogen bonds (rather than the three formed in a C-G pair)—strands rich in these nucleotides are generally easier to separate. Once strands are separated, RNA primers are created on the template strands and DNA polymerase extends these to create newly synthesized DNA. As DNA synthesis continues, the original DNA strands continue to unwind on each side of the bubble, forming replication forks. In bacteria, which have a single origin of replication on their circular chromosome, this process eventually creates a "theta structure" (resembling the Greek letter theta: è). In contrast, eukaryotes have longer linear chromosomes and initiate replication at multiple origins within these.

THE REPLICATION FORK

The replication fork is a structure which forms when DNA is being replicated. It is created through the action of helicase, which breaks the hydrogen bonds holding the two DNA strands together. The resulting structure has two branching "prongs", each one made up of a single strand of DNA.

Leading Strand Synthesis

In DNA replication, the leading strand is defined as the new DNA strand at the replication fork that is synthesized in the 5'"!3' direction in a continuous manner. When the enzyme helicase unwinds DNA, two single stranded regions of DNA (the "replication fork") form. On the leading strand DNA polymerase III is able to synthesize DNA using the free 3' OH group donated by a single RNA primer and continuous synthesis occurs in the direction in which the replication fork is moving.

Lagging Strand Synthesis

The lagging strand is the DNA strand at the opposite side of the replication fork from the leading strand, running in the 3' to 5' direction. Because DNA polymerase cannot synthesize in the 3'"!5' direction, the lagging strand is synthesized in short segments known as Okazaki fragments. Along the lagging strand's template, primase builds RNA primers in short bursts.

DNA polymerases are then able to use the free 3' OH groups on the RNA primers to synthesize DNA in the 5'"!3' direction. The RNA fragments are then removed (different mechanisms are used in eukaryotes and prokaryotes) and new deoxyribonucleotides are added to fill the gaps where the RNA was present. DNA ligase then joins the deoxyribonucleotides together, completing the synthesis of the lagging strand.

DYNAMICS AT THE REPLICATION FORK

As helicase unwinds DNA at the replication fork, the DNA ahead is forced to rotate. This process results in a build-up of twists in the DNA ahead. This build-up would form a resistance that would eventually halt the progress of the replication fork. DNA topoisomerases are enzymes that solve these physical problems in the coiling of DNA. Topoisomerase I cuts a single backbone on the DNA, enabling the strands to swivel around each other to remove the build-up of twists. Topoisomerase II cuts both backbones, enabling one double-stranded DNA to pass through another, thereby removing knots and entanglements that can form within and between DNA molecules.

Bare single-stranded DNA has a tendency to fold back upon itself and form secondary structures; these structures can interfere with the movement of DNA polymerase. To prevent this, single-strand binding proteins bind to the DNA until a second strand is synthesized, preventing secondary structure formation.

Clamp proteins form a sliding clamp around DNA, helping the DNA polymerase maintain contact with its template and thereby assisting with processivity. The inner face of the clamp enables DNA to be threaded through it. Once the polymerase reaches the end of the template or detects double stranded DNA, the sliding clamp undergoes a conformational change which releases the DNA polymerase. Clamp-loading proteins are used to initially load the clamp, recognizing the junction between template and RNA primers.

Eukaryotes

Within eukaryotes, DNA replication is controlled within the context of the cell cycle. As the cell grows and divides, it progresses through stages in the cell cycle; DNA replication occurs during the S phase (Synthesis phase). The progress of the eukaryotic cell through the cycle is controlled by cell cycle checkpoints. Progression through checkpoints is controlled through complex interactions between various proteins, including cyclins and cyclin-dependent kinases.

The G1/S checkpoint (or restriction checkpoint) regulates whether eukaryotic cells enter the process of DNA replication and subsequent division. Cells which do not proceed through this checkpoint are quiescent in the "G0" stage and do not replicate their DNA.

Replication of chloroplast and mitochondrial genomes occurs independent of the cell cycle, through the process of D-loop replication.

Bacteria

Most bacteria do not go through a well-defined cell cycle and instead continuously copy their DNA; during rapid growth this can result in multiple rounds of replication occurring concurrently. Within *E coli,* the most well-characterized bacteria, regulation of DNA replication can be achieved through several mechanisms, including: the hemimethylation and sequestering of the origin sequence, the ratio of ATP to ADP, and the levels of protein DnaA. These all control the process of initiator proteins binding to the origin sequences.

Because *E coli* methylates GATC DNA sequences, DNA synthesis results in hemimethylated sequences. This hemimethylated DNA is recognized by a protein (SeqA) which binds and sequesters the origin sequence; in addition, dnaA (required for initiation of replication) binds less well to hemimethylated DNA. As a result, newly replicated origins are prevented from immediately initiating another round of DNA replication. ATP builds up when the cell is in a rich medium, triggering DNA replication once the cell has reached a specific size. ATP competes with ADP to bind to DnaA, and the DNA-ATP complex is able to initiate replication. A certain number of DnaA proteins are also required for DNA replication — each time the origin is copied the number of binding sites for DnaA doubles, requiring the synthesis of more DnaA to enable another initiation of replication.

TERMINATION OF REPLICATION

Because bacteria have circular chromosomes, termination

of replication occurs when the two replication forks meet each other on the opposite end of the parental chromosome. *E coli* regulate this process through the use of termination sequences which, when bound by the Tus protein, enable only one direction of replication fork to pass through. As a result, the replication forks are constrained to always meet within the termination region of the chromosome.

Eukaryotes initiate DNA replication at multiple points in the chromosome, so replication forks meet and terminate at many points in the chromosome; these are not known to be regulated in any particular manner. Because eukaryotes have linear chromosomes, DNA replication often fails to synthesize to the very end of the chromosomes (telomeres), resulting in telomere shortening.

This is a normal process in somatic cells — cells are only able to divide a certain number of times before the DNA loss prevents further division. Within the germ cell line, which passes DNA to the next generation, the enzyme telomerase extends the repetitive sequences of the telomere region to prevent degradation. Telomerase can become mistakenly active in somatic cells, sometimes leading to cancer formation.

ROLLING CIRCLE REPLICATION

Another method of copying DNA, sometimes used *in vivo* by bacteria and viruses, is the process of rolling circle replication.

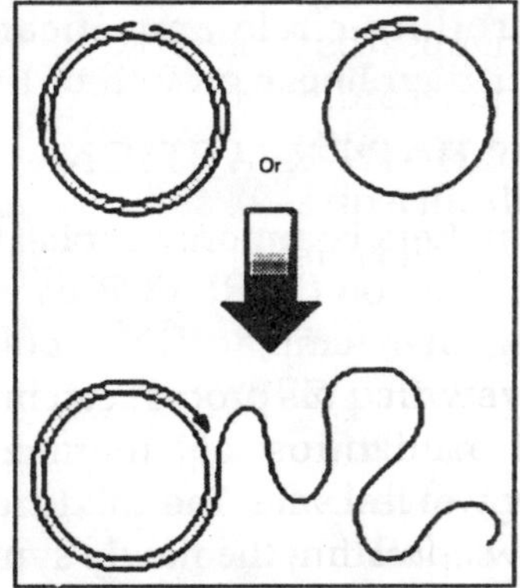

Fig. Rolling Circle Replication

In this form of replication, a single replication fork progresses around a circular molecule to form multiple linear copies of the DNA sequence. In cells, this process can be used to rapidly synthesize multiple copies of plasmids or viral genomes.

In the cell, rolling circle replication is initiated by an initiator protein encoded by the plasmid or virus DNA. This protein is able to nick one strand of the double-stranded, circular DNA molecule at a site called the double-strand origin (DSO) and remains bound to the 5' phosphate end of the nicked strand.

The free 3' hydroxyl end is released and can serve as a primer for DNA synthesis. Using the unnicked strand as a template, replication proceeds around the circular DNA molecule, displacing the nicked strand as single-stranded DNA. Continued DNA synthesis produces multiple single-stranded linear copies of the original DNA in a continuous head-to-tail series. *In vivo* these linear copies are subsequently converted to double-stranded circular molecules.

Rolling circle replication can also be performed *in vitro* and has found wide uses in academic research and biotechnology, often used for amplification of DNA from very small amounts of starting material. Replication can be initiated by nicking a double-stranded circular DNA molecule or by hybridizing a primer to a single-stranded circle of DNA. The use of a reverse primer (or random primers) produces hyperbranched rolling circle amplification, resulting in exponential rather than linear growth of the DNA molecule.

POLYMERASE CHAIN REACTION

In vitro, researchers commonly replicate DNA using the polymerase chain reaction (PCR). PCR uses a pair of primers to span a target region in template DNA, polymerizing partner strands in each direction. This process can be repeated through multiple cycles through the use of a thermostable polymerase.

At the start of each cycle, the mixture of template and primers is heated, separating the newly synthesized molecule and template. Then, as the mixture cools, both of these become templates for new primers to anneal to, and the polymerase extends from these. As a result the number of copies of the target region doubles each round, growing exponentially.

PROKARYOTIC DNA REPLICATION

Duplication of DNA is one of the fundamental properties of "molecule of life". Duplication or replication takes place once in a cell cycle. The duration and initiation point differs from one system to another. When conditions are favorable cell cytoplasmic mass increases and when cytoplasmic mass reaches a ratio to that of the cell size to 2L, bacterial cell division is triggered.

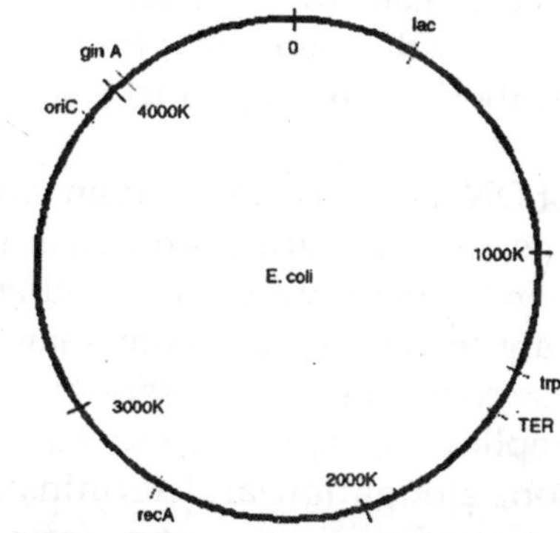

Fig. The Basic Circular Model of the Chromosomes with Map Positions such as Ori C and TER Regions

To complete its cell cycle it requires hardly 30 -40 minutes. Several genes involved in different steps of cell division have been characterized. Par excellent system to understand molecular process of prokaryotic cell division is E.coli. The size of the E.coli chromosome is 4.6×10^6 base pairs, but it is compacted by the binding of histone like proteins.

First the nucleoid i.e the chromosome frees from the membrane and unwinds. Such DNA initiates replication and completes in a matter of 15-20 minutes. Then the cytoplasm starts dividing almost in the centre of the cell by central septal ring formation.

It is at this time the daughter DNA molecules are partitioned and segregated to the two compartments. The central periseptal ring that started 1-2 minutes after DNA replication now furrows inwards and divides the cytoplasm in the middle region. Then mid-wall splits in the middle and daughter cell are set free, hence the Schizomycetes to bacteria (E.coli).

Bacterial cell division requires the participation of the following genes and gene products. Eukaryotic cell division is more complex and highly regulated. The paradigm for it is yeast, for it is a unicellular system, grows fast and requires hardly 60 minutes for one cell cycle. One can obtain different kinds of mutation like conditional and null or auxotrophic kinds. In culture cells, the duration of one cell cycle is about 12 hrs. Onion root tip cells divide once in 12 hrs. Whereas embryonic cells in its early stages require just 30 minutes for one cell cycle and they go through 6 to 8 such cell divisions very fast.

Replication of DNA in cell division is an important phase. The replication is very accurate and it is nucleotide by nucleotide and there is no provision for making any mistakes, even if mistakes are made they are immediately corrected or repaired, otherwise consequences are serious and deleterious. In general DNA replication is precise, exact and regulated and involves initiation, elongation and termination steps, but mechanisms and rate of replication vary from one system to the other.

REPLICATION IS SEMI CONSERVATIVE

Meselson and Stahl showed that DNA replicates in semi conservative mode. There were some doubts about the mode whether it is conservative, dispersive or semi conservative. But Meselson and Stahl used nonradioactive heavy Nitrogen isotope (15N), as the source of Nitrogen. When cells were grown on such source, 15^N incorporated nitrogen bases get incorporated into DNA, thus it becomes heavier than the DNA that is grown in normal (14N) nitrogen media. When two such DNAs, one grown in the presence of 15N source and the other in 14N (normal), are extracted separately and stained with ethidium Bromide, then subjected to equilibrium density centrifugation at ultra speed of 42000rpm for about 24 –36 hrs, the DNA, separates as two distinct bands, the one that is heavier (15N) is the lower band than the normal (14N) upper DNA.

If cells grown in heavy isotopes are shifted to normal Nitrogen source and grown for one or two cell generation and if DNA is isolated from it and if one combine this with the first two DNA sources and subject them to equilibrium density ultra centrifugation (using cesium chloride or cesium sulphate), one can observe three bands, the top DNA band is from cells grown in 14N without isotope, the second band is the hybrid band with one strand 15^N and the other with normal 14N-Nitrogen and the third band is with heavy nitrogen where both the strands are labeled with 15N.

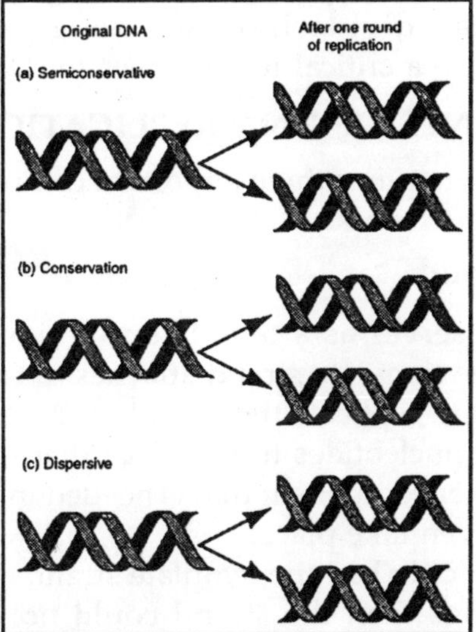

This experiment virtually and unambiguously proved that DNA replicates in semi conservative mode. Taylor using root tips, labeled with 32^P radioisotopes, of Vicia faba demonstrated semi conservative mode of replication at chromosomal level.

Semi conservative means that the two daughter molecules produced at the end of replication, each of the molecules retains one parental strand and the other one is the newly made one.

So it is semi conservative. The mode of replication is absolutely semi conservative irrespective the kind of DNA, whether it is ds DNA or ssDNA, whether it is linear or circular, and the basic mechanism is more or less similar; each of them uses their own specialized mechanisms. Experimentally semiconservative mode of replication has been shown by density gradient centrifugation of non-radioactive isotopes such as 15N and 14N sources.

DNA REPLICATION

DNA Replication (also known as DNA synthesis) is a process where the double stranded Deoxyribonucleic Acid (DNA) is copied. Replication is an important first step in cell division, as cells must duplicate their entire genetic constitution before they can divide into two daughter cells. DNA replication is also a critical requirement for DNA repair.

REQUIREMENTS OF DNA REPLICATION

Replication requires three important components in order for this process to work.

Template Strand

The DNA serves as a template to guide the incoming nucleotides. The template strand attaches to the RNA primer strand in order to replicate the strand of DNA. The template just guides the nucleotides to the place that they need to go in. This is another component that is needed in order for DNA replication to even take place.

If the DNA only had the template strand and none of the rest of the things then the strand could not replicate. The template starnd is very thin. Through an enzyme called primase the DNA template strand is synthesized one nucleotide at a time.

DNA Polymerase

There are many DNA polymerases within a cell, but only one of which is used in the replication of DNA. In order for DNA to replicate, it must have a primer, which is a small strand of RNA. The DNA template strand synthesizes one nucleotide at a time by the use of primase. The adding of nucleotides to the 3′ end, which is continued until that section of DNA is replicated.

The other DNA polymerases that are within the cell are used in other ways besides replication. DNA polymerase III is one certain polymerase, which aids in replication of what is known as the "lagging strand". DNA ligase is the final enzyme, which combines the lagging strand

Free 3' Hydroxyl

The free 3' hydroxyl is the starter strand called a primer which is required in order for DNA to be replicated. The primer strand is not very big, it is shorter then a single strand of RNA. The primer is the complementary strand to the template strand of DNA. This is another one of the things that are needed in order to make the process of DNA replication even possible. The DNA polymerase is added to the 3' end of the primer, which keeps growing until the section is finally complete. After a while the RNA primer is degraded and disappears.

Types of Replication

During the discovery of DNA replication they also discovered that there are three modes of replication that a DNA strand can take.

The three types of DNA replication are semiconservative, conservative, and dispersive. These three types of replication were tested on in order to see the possible patterns that would result in the complementary base pairing.

Semiconservative Replication

Through this type of replication the parent strand serves as a template for the new strand. The two offsprings would have one of the parent strand and one new strand.

Conservative Replication

Through this type of replication the double helix serves

as a template. This although does not contribute to the new double helix.

Dispersive Replication

In this type of replication the fragments from the parent DNA molecule. The DNA serves as a template for the assembly of the new molecule. The new double helix contains old and new parts of the DNA strands.

DNA Repairing

After the DNA is replicated, there are three DNA repair mechanisms. They are to be used when they feel necessary. This will lessen the chance of human mutation. The rate of error in which the DNA polymerases makes a mistake is very high. This is why the DNA has mechanisms to cause fewer mutations. The three mechanisms are:

Proofreading

Every time a nucleotide is introduced to an already growing chain, the DNA polymerase checks the connection. If the pair is mismatched, it will be removed and redone. This lowers the overall rate of mutations.

Mismatch Pair

After the replication takes place, another set of proteins check to make sure that there are no more mismatched base pairs and also which strand is the wrong one.

Excision Pair

There can still be damage while the cell is living. Because of this, there are enzymes constantly checking the cell. If they detect a problem, the enzyme cuts the strand and also cuts away the opposite base. DNA polymerase and ligase then fix this base sequence.

LABORATORY TECHNIQUE

Scientists use two techniques in a laboratory that involve DNA replication. They use these techniques to observe genes and genomes. One of the techniques can make multiple copies of DNA from only a short piece. Another technique allows scientists to determine the base sequence of DNA. The two techniques are:

Polymerase Chain Reaction (PCR)

Through this process, a short strand of DNA is copied repetitively. There are three steps that are repeated in the process. The DNA strands are denatured or heated in order to separate the strand. Next, they add a primer, which was artificially made. They also add DNA polymerase and the four deoxyribonucleotide triphosphates. These three are needed in order to replicate DNA. The final step is then the DNA polymerase catalyzes the production of a complementary strand.

DNA Sequencing

This allows scientists to determines the base sequence of a DNA. The process is the DNA is first denatured and a single strand is placed in a test tube. They then add DNA polymerase, primer, the four dNTP's, and small amounts of ddNTP's. The DNA replicates within the tube and after the DNA fragments are denatured. The fragments then go through electrophoresis, which sorts the lengths of DNA fragments. The fragments then pass through a laser beam which excites the fluorescent tags. This information is fed into a computer at which tells the sequence.

Mendelson- Stahl Experiment

Through this experiment Meselson- Stahl convinced the scientists that the correct model for DNA replication was the semiconservative replication. They used density labeling in order to distinguish the old and new strands of DNA. In their experiment they used "heavy" isotopes of nitrogen. This nitrogen is a nonradioactive isotope that made the molecule more dense. The chemically identical molecules containing an isotope known as ^{14}N. they needed a way to distinguish the different DNA densities.

Because of this they needed to come up with something

to measure the densities of solutions. Meselson, Stahl, and Jerome Vanguard came up with centrifuge, which is a procedure that involves the spinning of solutions at high speed, which causes the particles to separate and form gradient according to the different densities of the solution. The first part of their experiment they grew cultures of Escherichia coli. One of the cultures was grown in ^{15} N, which made the DNA "heavy".

The other culture was placed in a medium using ^{14}N rather than ^{15}N, which made all the DNA "light". In order to see if the use of centrifuge worked they combined the two cultures and when centrifuged it formed two DNA bands. This showed them the different densities with in the two cultures. The next part of their experiment they grew E. Coli in the ^{15}N medium and then transferred the bacteria from that medium to the ^{14}N. Through this they came up with the conclusion that through E. Coli DNA replicates every 20 minutes. They then took DNA samples from each generation through, which they found that the density gradient was different each generation.

Their observations could only be explained with the semiconservative model of DNA replication. Their results showed that when the DNNA first underwent replication, it was in 15N, which caused the DNA to be heavy. Because in the semiconservative model of DNA replication, the one strand acts as a template for a second strand which the DNA was in a ^{14}N DNA strand and a ^{15}N and they were of intermediate density.

Chapter 7

Regulation of Gene Expression

Regulation of gene expression (or gene regulation) refers to the cellular control of the amount and timing of changes to the appearance of the functional product of a gene. Although a functional gene product may be an RNA or a protein, the majority of known mechanisms regulate protein coding genes. Any step of the gene's expression may be modulated, from DNA-RNA transcription to the post-translational modification of a protein.

Gene regulation is essential for viruses, prokaryotes and eukaryotes as it increases the versatility and adaptability of an organism by allowing the cell to express protein when needed. The first example of gene regulation system was the lac operon, discovered by Jacques Monod, in which protein involved in lactose metabolism are expressed by E.coli only in the presence of lactose and absence of glucose.

Furthermore, gene regulation allows the presence in a multicellular organism of different cells types arranged in a complex pattern, hence different transcriptomes despite them having all the same genome and the generation of patterns by cellular differentiation and morphogenesis.

REGULATED STAGES OF GENE EXPRESSION

Any step of gene expression may be modulated, from the DNA-RNA transcription step to post-translational modification of a protein. The following is a list of stages where gene expression is regulated:

- Chromatin domains
- Transcription

- Post-transcriptional modification
- RNA transport
- Translation
- mRNA degradation
- Post-translational modifications

MODIFICATION OF DNA

In eukaryotes, the accessibility of large regions of DNA can depend on its chromatin structure which can be altered as a result of histone modifications which are directed by DNA methylation, ncRNA or DNA binding protein.

Chemical

Methylation of DNA is a common method of gene silencing. DNA is typically methylated by methyltransferase enzymes on cytosine nucleotides in a CpG dinucleotide sequence (also called "CpG islands" when densely clustered). Analysis of the pattern of methylation in a given region of DNA (which can be a promoter) can be achieved through a method called bisulfite mapping.

Methylated cytosine residues are unchanged by the treatment, whereas unmethylated ones are changed to uracil. The differences are analyzed by DNA sequencing or by methods developed to quantify SNPs, such as Pyrosequencing (Biotage) or MassArray (Sequenom), measuring the relative amounts of C/T at the CG dinucleotide. Abnormal methylation patterns are thought to be involved in carcinogenesis.

Structural

Transcription of DNA is dictated by its structure. In general, the density of its packing is indicative of the frequency of transcription. Octameric protein complexes called histones are responsible for the amount of supercoiling of DNA, and these complexes can be temporarily modified by processes such as phosphorylation or more permanently modified by processes such as methylation. Such modifications are considered to be responsible for more or less permanent changes in gene expression levels.

Histone acetylation is also an important process in transcription. Histone acetyltransferase enzymes (HATs) such as CREB-binding protein also dissociate the DNA from the histone complex, allowing transcription to proceed. Often, DNA methylation and histone deacetylation work together in gene silencing. The combination of the two seems to be a signal for DNA to be packed more densely, lowering gene expression.

REGULATION OF TRANSCRIPTION

Regulation of transcription controls when transcription occurs and how much RNA is created. Transcription of a gene by RNA polymerase can be regulated by at least five mechanisms:

- *Specificity factors* alter the specificity of RNA polymerase for a given promoter or set of promoters, making it more or less likely to bind to them (i.e. sigma factors used in prokaryotic transcription).
- *Repressors* bind to non-coding sequences on the DNA strand that are close to or overlapping the promoter region, impeding RNA polymerase's progress along the strand, thus impeding the expression of the gene.
- *General transcription factors* These transcription factors position RNA polymerase at the start of a protein-coding sequence and then release the polymerase to transcribe the mRNA.
- *Activators* enhance the interaction between RNA polymerase and a particular promoter, encouraging the expression of the gene. Activators do this by increasing the attraction of RNA polymerase for the promoter, through interactions with subunits of the RNA polymerase or indirectly by changing the structure of the DNA.
- *Enhancers* are sites on the DNA helix that are bound to by activators in order to loop the DNA bringing a specific promoter to the initiation complex.

POSTTRANSCRIPTIONAL REGULATION

After the DNA is transcribed and mRNA is formed there must be some sort of regulation on how much the mRNA is translated into Proteins. Cells do this by modulating the Capping, Splicing, addition of a Poly(A) Tail, the sequence-specific nuclear export rates and in several contexts sequestration of the RNA transcript. These processes occur in eukaryotes but not in prokaryotes. This modulation is a result of a protein or transcript which in turn is regulated and may have an affinity for certain sequences.

- *Capping* changes the five prime end of the mRNA to a three prime end by 5'-5' linkage, which protects the mRNA from 5' exonuclease, which degrades foreign RNA. The cap also helps in ribosomal binding.
- *Splicing* removes the introns, noncoding regions that are transcribed into RNA, in order to make the mRNA able to create proteins. Cells do this by spliceosome's binding on either side of an intron, looping the intron into a circle and then cleaving it off. The two ends of the exons are then joined together.
- *Addition of poly(A) tai*l otherwise known as poly-adenylation. Junk RNA is added to the 3' end, and acts as a buffer to the 3' exonuclease in order to increase the half life of mRNA.

In both prokaryotes and eukaryotes a large number of RNA binding protein exist, with often are directed to their target sequence by the secondary structure of the transcript, which may change depending on certain conditions, such as temperature or presence of a ligand (aptamer), some transcripts act as ribozymes and self-regulate their expression.

EXAMPLES OF GENE REGULATION

- Enzyme induction is a process in which a molecule (e.g. a drug) induces (i.e. initiates or enhances) the expression of an enzyme.

- The induction of heat shock proteins in the fruit fly *Drosophila melanogaster*.
- The Lac operon is an interesting example of how gene expression can be regulated.
- Viruses despite having only a few genes, possess mechanisms to regulate their gene expression, typically into a early and late phase, using collinear systems regulated by anti-terminators (lambda phage) or splicing modulators (HIV)

CIRCUITRY

UP-REGULATION AND DOWN-REGULATION

Up-regulation is a process which occurs within a cell triggered by a signal (originating internal or external to the cell) which results in increased expression of one or more genes and as a result the protein(s) encoded by those genes. Conversely down-regulation is a process resulting in decreased gene and corresponding protein expression.

- Up: Regulation occurs for example when a cell is deficient in some kind of receptor. In this case, more receptor protein is synthesized and transported to the membrane of the cell and thus the sensitivity of the cell is brought back to normal reestablishing homeostasis.
- Down: Regulation occurs for example when a cell is overly stimulated by a neurotransmitter, hormone, or drug for a prolonged period of time and the expression of the receptor protein is decreased in order to protect the cell.

INDUCIBLE VS. REPRESSIBLE SYSTEMS

Gene Regulation can be summarized as how they respond:

- *Inducible systems:* An inducible system is off unless there is the presence of some molecule (called an inducer) that allows for gene expression. The molecule is said to "induce expression". The manner in which this happens is dependent on the control mechanisms as well as differences between prokaryotic and eukaryotic cells.

- *Repressible Systems:* A repressible system is on except in the presence of some molecule (called a corepressor) that suppresses gene expression. The molecule is said to "repress expression". The manner in which this happens is dependent on the control mechanisms as well as differences between prokaryotic and eukaryotic cells.

DEVELOPMENTAL BIOLOGY

A large number of studied regulatory systems come from developmental biology. Examples include:

- The collinearity of the Hox gene cluster with their nested antero-posterior patterning
- It has been speculated that pattern generation of the hand (digits - interdigits) The gradient of Sonic hedgehog (secreted inducing factor) from the zone of polarizing activity in the limb which creates a gradient of active Gli3 which activates Gremlin which inhibits BMPs also secreted in the limb resulting in the formation of an alternating pattern of activity as a result of this reaction-diffusion system.
- Somatogenesis is the creation of segmatation (somites) form a uniform tissue (PSM) sequentially from anterior to posterior, this is a achieved in amniotes possibly by means of two opposing gradients, Retinoic acid in the anterior (wavefront) and an oscillating gradient in the posterior (clock) composed of FGF + Notch and Wnt in antiphase.
- Sex determination in the soma of a Drosophila requires the sensing of the ratio of autosomal genes to sex chromosome encoded genes, which results in the production of sexless splicing factor in females resulting in the female isoform of doublesex.

THEORETICAL CIRCUITS

- Repressor/Inducer: a activation of a sensor results in the change of expression of a gene

- Negative feedback: the gene product downregulates its own production directly or indirectly, which can result in
  - Keeping transcript levels constant/proportional to a factor
  - Inhibition of run-away reactions when coupled with a positive feedback loop
  - Creating an oscillator by taking advantage in the time delay of transcription and translation, given that the mRNA and protein half-life is shorter
- Positive feedback: the gene product upregulates its own production directly or indirectly, which can result in
  - Signal amplification
  - Bistable switches when two genes inhibit each other and have both positive feedback
  - Pattern generation

Methods

Generally, most experiments investigating differential expression used whole cell extracts of RNA, called steady-state levels, to determined which genes changed and by how much they did, these are however not informative of where the regulation has occurred and may actually mask conflicting regulatory processess, it is the most commonly analysed (QPCR and DNA microarray).

When studying gene expression there are several methods to look at the various stages. In eukaryotes these include:

- The chromatin conformation of the region can be determined by ChIP-chip analysis by pulling down RNA Polymerase II, Histone 3 modifications, Trithorax-group protein,Polycomb-group protein or any other DNA binding element to which a good antibody is available.
- Due to post-transcriptional regulation, transcription rates and total RNA levels differ significantly, to measure the transcription rates nuclear run-on

assays can be done and newer high-throughput methods are being developed, using thiol labelling instead of radioactivity.

- Only 5% of the RNA polymerised in the nucleus actually exists and not only introns, abortive products and non-sense transcripts are degradated therefore the differences in nuclear and cytoplasmic levels can be see by separating the two fractions by gentle lysis.
- Alternative splicing can be analysed with a splicing array or with a tiling array.
- All in vivo RNA is complexed as RNPs. The quantity of transcripts bound to specific protein can be also analysed by RIP-Chip, for example DCP2 will give an indication of sequestered protein, ribosome bound gives and indication of transcripts active in transcription (although it should be noted that a more dated method, called polysome fractionation, is still popular in some labs)
- Protein levels can be analysed by Mass spectrometry, which can only be compare to QPCR data as microarray data is relative and not absolute.
- RNA and protein degradation rates are measured by means of transcription inhibitors (actinomycin D or á-amanitin) or translation inhibitors (Cycloheximide) respectively.

GENE EXPRESSION

Gene expression is the process by which inheritable information from a gene, such as the DNA sequence, is made into a functional gene product, such as protein or RNA.

Several steps in the gene expression process may be modulated, including the transcription step and the post-translational modification of a protein. Gene regulation gives the cell control over structure and function, and is the basis for cellular differentiation, morphogenesis and the versatility and adaptability of any organism. Gene regulation may also serve as a substrate for evolutionary change, since control of the timing, location, and amount of gene expression can have a profound effect on the functions (actions) of the gene in the organism.

Non-protein coding genes (e.g. rRNA genes, tRNA genes) are transcribed, but not translated into protein.

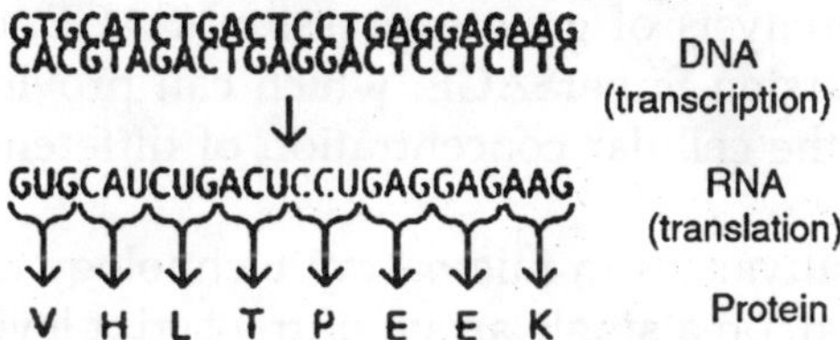

Genes are expressed by being transcribed into RNA, and this transcript may then be translated into protein.

MEASUREMENT

The expression of many genes is regulated after transcription (i.e., by microRNAs or ubiquitin ligases), so an increase in mRNA concentration need not always increase expression. Nevertheless, mRNA levels can be quantitatively measured by Northern blotting, a process in which a sample of RNA is separated on an agarose gel and hybridized to a radio-labeled RNA probe that is complementary to the target sequence.

Northern blotting requires the use of radioactive reagents and can have lower data quality than more modern methods (due to the fact that quantification is done by measuring band strength in an image of a gel), but it is still often used. It does, for example, offer the benefit of allowing the discrimination of alternately spliced transcripts.

A more modern low-throughput approach for measuring mRNA abundance is real-time polymerase chain reaction (The term RT-PCR is used to refer to both reverse transcription PCR as well as real-time PCR, which is also known as quantitative RT-PCR or quantitative PCR (qPCR).

With a carefully constructed standard curve qPCR can produce an absolute measurement such as number of copies of mRNA per nanolitre of homogenized tissue. The lower level of noise in data obtained via qPCR often makes this the method of choice, but the price of the required equipment and reagents can be prohibitive.

In addition to low-throughput methods, transcript levels for many genes at once (expression profiling) can be measured with DNA microarray technology or "tag based" technologies like Serial analysis of gene expression (SAGE) or the more advanced version SuperSAGE, which can provide a relative measure of the cellular concentration of different messenger RNAs.

Recent advances in microarray technology allow for the quantification, on a single array, of transcript levels for every known gene in the human genome. The great advantage of tag-based methods is the "open architecture", allowing for the exact measurement of any transcript, known or unknown. Especially SuperSAGE recommends itself therefore also for studying organisms with unknown genomes.

Protein levels themselves can be estimated by a number of means. The most commonly used method is to perform a Western blot against the protein of interest, whereby cellular lysate is separated on a polyacrylamide gel and then probed with an antibody to the protein of interest.

The antibody can either be conjugated to a fluorophore or to horseradish peroxidase for imaging or quantification. Another commonly used method for assaying the amount of a particular protein in a cell is to fuse a copy of the protein to a reporter gene such as Green fluorescent protein, which can be directly imaged using a fluorescent microscope.

Because it is very difficult to clone a GFP-fused protein into its native location in the genome, however, this method often cannot be used to measure endogenous regulatory mechanisms (GFP-fusions are therefore most often expressed on extra-genomic DNA such as an expression vector). Fusing a target protein to a reporter can also change the protein's behaviour, including its cellular localization and expression level.

The pattern of detection of a gene or gene product may be described using terms such as facultative, constitutive, circadian, cyclic, housekeeping, or inducible.

REGULATION OF GENE EXPRESSION

Regulation of gene expression is the cellular control of the amount and timing of appearance of the functional product of a gene. Any step of gene expression may be modulated, from the DNA-RNA transcription step to post-translational modification of a protein. Gene regulation gives the cell control over structure and function, and is the basis for cellular differentiation, morphogenesis and the versatility and adaptability of any organism.

Expression System

An expression system consists, minimally, of a source of DNA and the molecular machinery required to transcribe the DNA into mRNA and translate the mRNA into protein using the nutrients and fuel provided. In the broadest sense, this includes every living cell capable of producing protein from DNA.

However, an expression system more specifically refers to a laboratory tool, often artificial in some manner, used for assembling the product of a specific gene or genes. It is defined as the "combination of an expression vector, its cloned DNA, and the host for the vector that provide a context to allow foreign gene function in a host cell, that is, produce proteins at a high level".

In addition to these biological tools, certain naturally observed configurations of DNA (genes, promoters, enhancers, repressors) and the associated machinery itself are referred to as an expression system, as in the simple repressor 'switch' expression system in Lambda phage. It is these natural expression systems that inspire artificial expression systems, (such as the Tet-on and Tet-off expression systems).

Each expression system has distinct advantages and liabilities, and may be named after the host, the DNA source or the delivery mechanism for the genetic material. For example, common expression systems include bacteria (such as E.coli), yeast (such as S.cerevisiae), plasmid, artificial chromosomes, phage (such as lambda), cell lines, or virus (such as baculovirus, retrovirus, adenovirus).

Overexpression

In the laboratory, the protein encoded by a gene is sometimes expressed in increased quantity. This can come about by increasing the number of copies of the gene or increasing the binding strength of the promoter region.

Often, the DNA sequence for a protein of interest will be cloned or subcloned into a plasmid containing the *lac* promoter, which is then transformed into the bacterium *Escherichia coli*. Addition of IPTG (a lactose analog) causes the bacteria to express the protein of interest. However, this strategy does not always yield functional protein, in which case, other organisms or tissue cultures may be more effective.

For example, the yeast *Saccharomyces cerevisiae* is often preferred to bacteria for proteins that undergo extensive posttranslational modification. Nonetheless, bacterial expression has the advantage of easily producing large amounts of protein, which is required for X-ray crystallography or nuclear magnetic resonance experiments for structure determination.

GENE NETWORKS AND EXPRESSION

Genes have sometimes been regarded as nodes in a network, with inputs being proteins such as transcription factors, and outputs being the level of gene expression. The node itself performs a function, and the operation of these functions have been interpreted as performing a kind of information processing within cell and determine cellular behaviour.

CONTROL OF GENE EXPRESSION

While the period from 1900 to the second world war has been called the "golden age of genetics", we may be in a new golden (or platinum) age. Recombinant DNA technology allows us to manipulate the very DNA of living organisms and to make conscious changes in that DNA. Prokaryote genetic systems are much easier to study and better understood than are eukaryote systems.

The Chromosome of *E. Coli*

The single chromosome of the common intestinal bacterium *E. coli* is circular and contains some 4.7 million base pairs. It is nearly 1 mm long, but only 2nm wide. The chromosome replicates in a bidirectional method, producing a figure resembling the Greek letter theta. The promoter is the part of the DNA to which the RNA polymerase binds before opening the segment of the DNA to be transcribed.

A segment of the DNA that codes for a specific polypeptide is known as a structural gene. These often occur together on a bacterial chromosome. The location of the polypeptides, which may be enzymes involved in a biochemical pathway, for example, allows for quick, efficient transcription of the mRNAs. Often leader and trailer sequences, which are not translated, occur at the beginning and end of the region. *E. coli* can synthesize 1700 enzymes. Therefore, this small bacterium has the genes for 1700 different mRNAs.

Lactose, milk sugar, is split by the enzyme b-galactosidase. This enzyme is inducible, since it occurs in large quantities only when lactose, the substrate on which it operates, is present. Conversely, the enzymes for the amino acid tryptophan are produced continuously in growing cells unless tryptophan is present. If tryptophan is present the production of tryptophan-synthesizing enzymes is repressed.

THE OPERON MODEL

The operon model of prokaryotic gene regulation was proposed by Fancois Jacob and Jacques Monod. Groups of genes coding for related proteins are arranged in units known as operons. An operon consists of an operator, promoter, regulator, and structural genes.

The regulator gene codes for a repressor protein that binds to the operator, obstructing the promoter (thus, transcription) of the structural genes. The regulator does not have to be adjacent to other genes in the operon. If the repressor protein is removed, transcription may occur. Operons are either inducible or repressible according to the control mechanism. Seventy-five different operons controlling 250 structural genes have been identified for *E. coli*. Both repression and induction are examples of negative control since the repressor proteins turn off transcription.

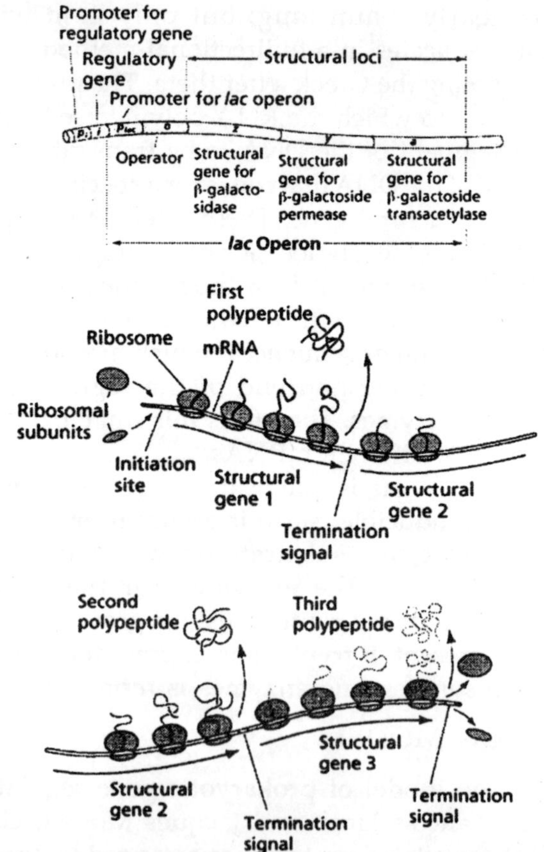

Fig. The Structure and Operation of an Operon

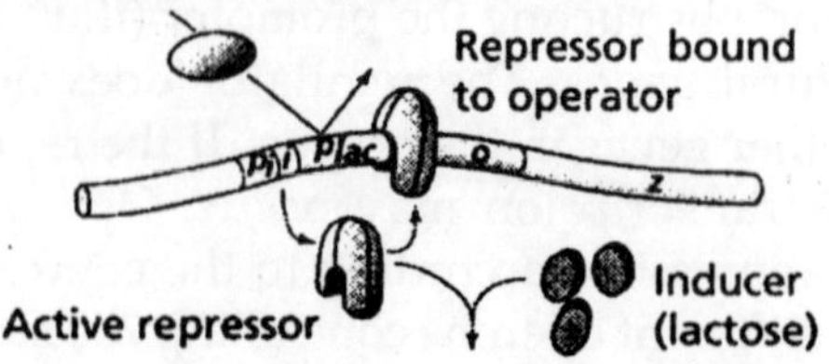

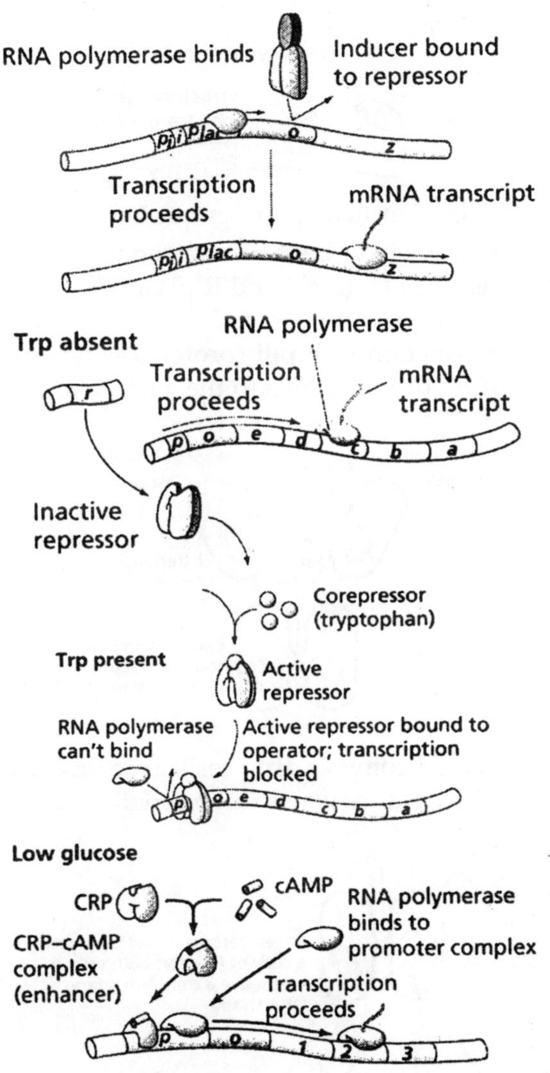

Plasmids, small DNA fragments, are known from almost all bacterial cells. Plasmids carry between 2 and 30 genes. Some seem to have the ability to move in and out of the bacterial chromosome. An episome is a plasmid incorporated in the bacterial chromosome. Plasmids are self-replicating in a manner like the bacterial chromosome.

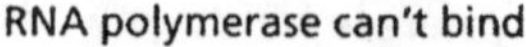

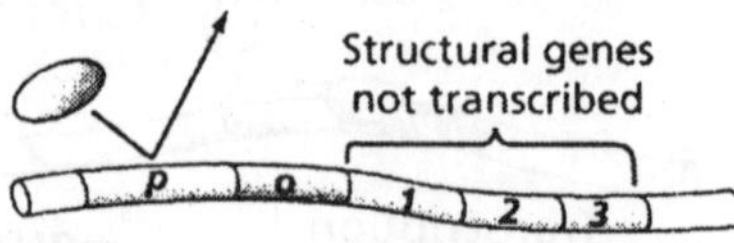

Fig. Functioning of Several Operons

Many plasmids have been recognized for *E. coli,* including the F plasmids ("sex factors") and R plasmids (drug/antibiotic resistance). The F plasmid contains 25 genes, some of which control the production of F pili (proteins which extend from the surface of F^+, or male, cells to the surface of F^-, or female, cells).

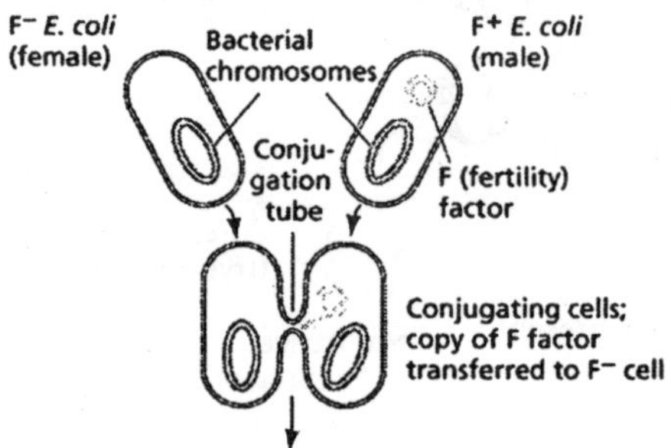

The R plasmid conveys drug resistance on cells having it. As many as 10 resistance genes can be contained on a single R plasmid.

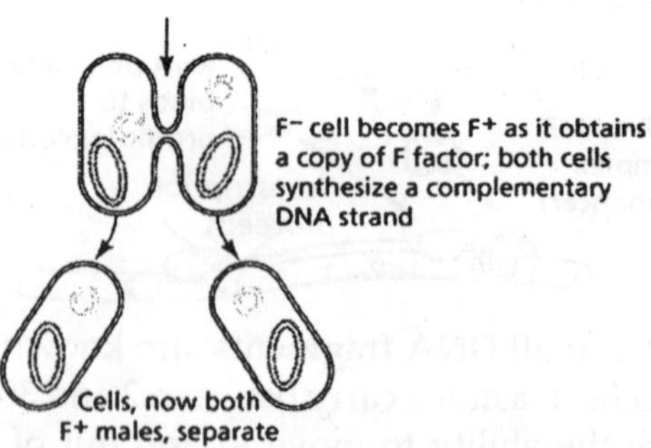

The R plasmids can be transferred to other bacteria of the same species, to viruses, and even to bacteria of different species. Drug (antibiotic) resistance has been found among pathogens causing the diseases typhoid fever, gastroenteritus, plague, undulant fever, meningitis, and gonorrhea.

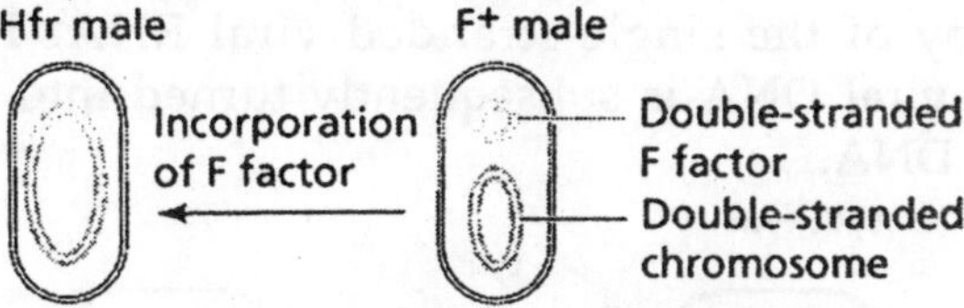

Fig. Conjugation and Exchange of Genetic Material in Bacteria

In addition to the more common modes of transfer, R plasmids may be passed through the cell membrane. The resistance genes appear to operate by either breaking down the antibiotics or by circumventing the block the antibiotic places on a key bacterial metabolic pathway.

VIRUSES

Viruses consist of a nucleic acid (DNA or RNA) enclosed in a protein coat (known as a capsid). The capsid may be a single protein repeated over and over, as in tobacco mosaic virus (TMV). It may also be several different proteins, as in the T-even bacteriophages.

Once inside the cell, the nucleic acid follows one of two paths: lytic or lysogenic.

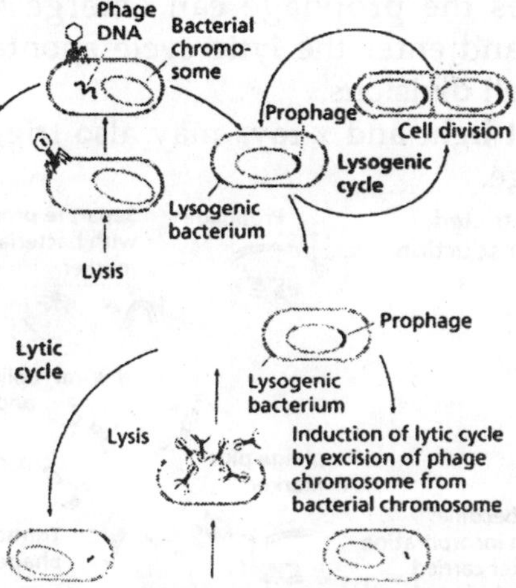

Retroviruses, such as Human Immunodifficiency Virus

(HIV), also include the enzyme reverse transcriptase with the viral RNA. Reverse transcriptase makes a single-stranded viral DNA copy of the single-stranded viral RNA. The single stranded viral DNA is subsequently turned into a double-stranded DNA.

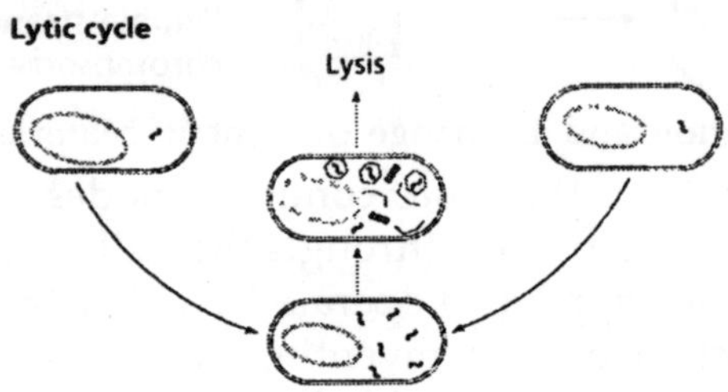

Fig. The Lytic and Lysogenic Phases of a Viral Replication Cycle

The lytic cycle occurs when the viral DNA immediately takes over the host cell (remember that viruses are obligate intracellular parasites) and begins making new viruses. Eventually the new viruses cause the rupture (or lysis) of the cell, releasing those new viruses to continue the infection cycle.

The lysogenic cycle occurs when the viral DNA is incorporated into the host DNA as a prophage. When the cell replicates the prophage is passed along as if it were host DNA.

Sometimes the prophage can emerge from the host chromosome and enter the lytic cycle spontaneously once every 10,000 cell divisions.

Ultraviolet light and x-rays may also trigger emergence of the prophage.

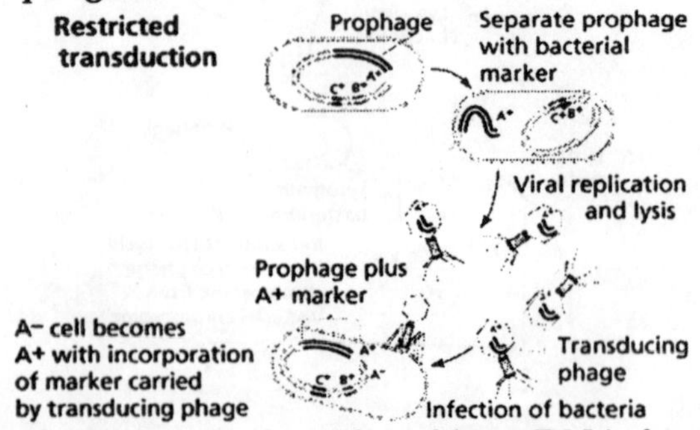

Transduction is the transfer of host DNA from one cell

to another by a virus. Some bacteriophages are temperate since they tend to go lysogenic rather than lytic. These types of viruses are able to transduce fragments of the host DNA. Transposons are DNA fragments incorporated into the chromosomal DNA. Unlike episomes and prophages, transposons contain a gene producing an enzyme that catalyzes insertion of the transposon at a new site.

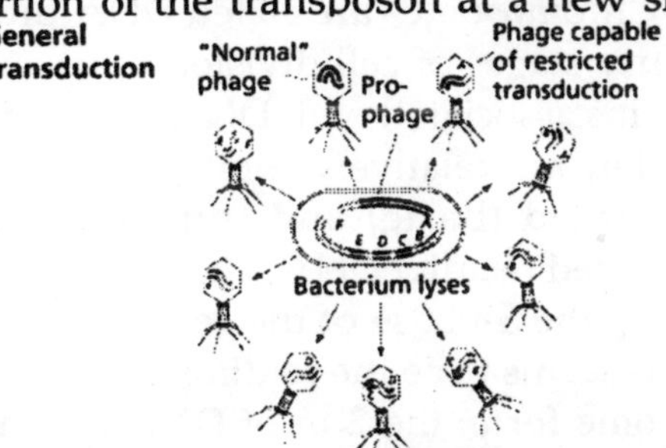

Fig. Induction of Transduction by Viruses in Bacteria

They also have repeated sequences 20-40 nucleotides in length at each end. Insertion sequences are short (600-1500 base pairs long) simple transposons that do not carry genes beyond those essential for insertion of the transposon into *E. coli*. Complex transposons are much larger and carry additional genes.

Genes incorporated in a complex transposon are known as jumping genes since they can move about on the chromosome (even from chromosome to chromosome). Often the complex transposons are flanked by simple transposons.

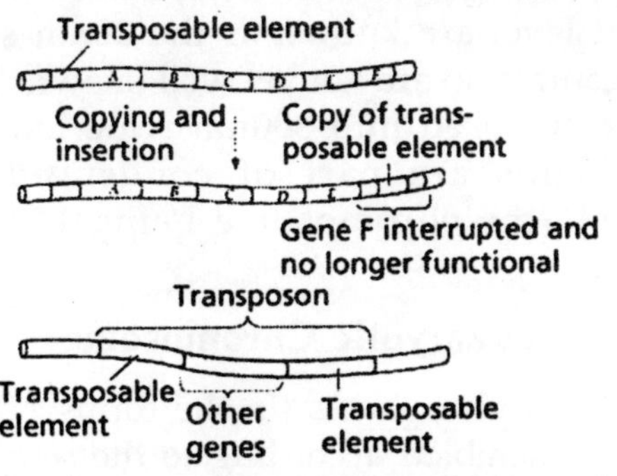

Fig. Transposons and their Relationship to other Genes

THE EUKARYOTIC CHROMOSOME

The eukaryotic chromosome consists of DNA and proteins that appear to play a major role in regulation of eukaryote genes. The DNA of each chromosome is a long single molecule of double stranded DNA. Eukaryotic DNA comes in two forms. Chromatin is the uncoiled form of DNA and is over 50% protein. Chromosomes are coiled DNA/protein that form during the early stages of cell division.

The proteins associated with DNA are collectively known as histones. They are relatively short polypeptides which are positively charged (basic) and thus are attracted to the negatively charged (acidic) DNA. Histones are synthesized in quantity during the S-phase of the cell cycle. One function of theses proteins seems to be the folding and packaging of DNA into chromosome form: the 2 m of DNA in a human cell are packaged into 46 chromosomes with a combined length of 200nm (a nm remember is 10-6m). Some 90 million molecules of histones occur in a single cell, with the majority (30 million) being H1 histones. Five types of histone are known (H1, H2A, H2B, H3, and H4); with the exception of H1 most eukaryote histones are very similar.

A nucleosome is the fundamental packing unit of eukaryotic DNA. The core consists of two molecules each of H2A, H2B, H3, and H4; around which the DNA is wound twice. The H1 histone is outside the core. Between 150-200 nucleotide pairs are associated with the core and linker DNA. This level of packing is known as "beads on a string".

The next level are known as the 30nm strand, whose details of organization are not yet well known. 30 nm strands are further condensed into 300nm wide looped domains. Looped domains are part of condensed sections of chromosomes (the chromosome being 1400nm wide at Metaphase I).

Replication of Eukaryotic Chromosome

Nucleotide triphosphates (in the forms ATP, GTP, CTP and TTP) are assembled according to the semi-conservative model. Other details of DNA replication are consistent with what we know for prokaryotes. Eukaryotic DNA has many replication forks and also bidirectional synthesis, contrasting to the unidirectional rolling theta prokaryotic method. Eukaryotic DNA is synthesized much slower than prokaryotic DNA, in humans 50 nucleotides per second per replication fork. After replication the new DNA is immediately associated with histones.

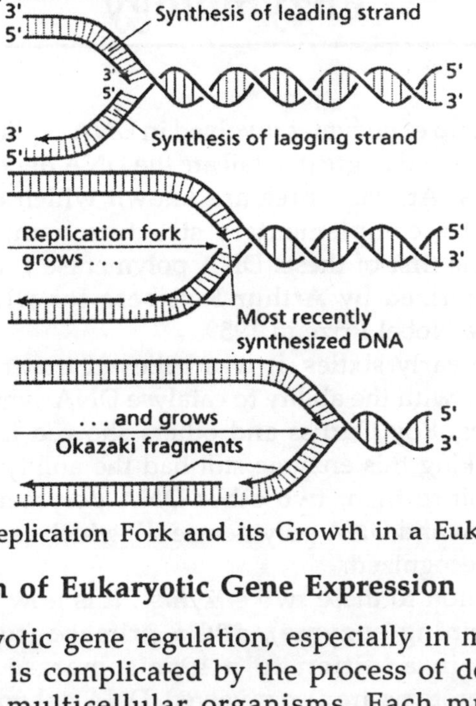

3'
5'
Synthesis of leading strand
3'
5'
5'
3'
3'
5'
Synthesis of lagging strand
Replication fork grows . . .
5'
3'
Most recently synthesized DNA
. . . and grows
Okazaki fragments
5'
3'

Fig. Replication Fork and its Growth in a Eukaryote.

Regulation of Eukaryotic Gene Expression

Eukaryotic gene regulation, especially in multicellular organisms, is complicated by the process of development unique to multicellular organisms. Each multicellular organism begins as a single-celled zygote which divides by mitosis. Cells differentiate into functional types by using some genes but ignoring others. Homeobox genes establish the body plan and position of organs in response to gradients of regulatory molecules. The timing of certain gene expressions seems to follow a sequence, such as the production of different types of fetal hemoglobins by mammalian red blood cells, which switch to adult hemoglobin sometime after birth. Clearly the inactivation of certain genes occurs in every adult cell; therein lies the cure for cancer, old age, etc.

Chapter 8

Enzymology

One group of enzymes involved in DNA replication which have been studied in great detail are the DNA dependent DNA polymerases. At least three are known which catalyse the synthesis of a complementary structure using DNA as a template. The first of these. DNA polymerase I, was isolated and characterized by Arthur Kornberg for which he was awarded the Nobel Prize in 1959.

Till the early sixties, it was believed that this was the only enzyme with the ability to catalyse DNA synthesis. Later in the sixties, Roy Curtiss and others showed that bacterial mutants !acking this enzyme still had the ability to replicate DNA and since then, two other DNA polymerases (DNA polymerase II and DNA polymerase III) which can copy DNA have been recognized.

In addition to these two enzymes, it is now known that a ligase (joining enzyme), and a primase (to prime the synthesis), in addition to a topoisomerase and other replication proteins are also involved. DNA polymerase I, was thought to be the major enzyme that joins together de-oxynucleotides in vivo but now it is realized that DNA polymerase III is the main polymerizing enzyme while DNA polymerase I is a repair enzyme and fills In the gaps between the small fragments. The role of DNA polymerase II is not yet clear.

All these three known polymerases extend the DNA molecules from the 5 end to the 3 end only. This led to the idea that the replication of double stranded DNA occurs in short pieces discontinuously and these short segments are later joined with the aid of a ligase to form a continuous strand. Also, the DNA polymerases can add nucleotides only to a perfectly base paired nucleotide sequence and they cannot initiate new DNA synthesis.

It now appears that the primary reaction is carried out by a DNA dependant RNA polymerase, which synthesizes short degradable RNA primers. It is also believed that in addition to the RNA polymerases, specific "primases", which differ from the RNA polymerases, synthesize the primary RNA which is then elongated by the DNA polymerase such a primase has been identified in the bacterium E. coli and also in the bacteriophages T4 and T7 Because of the difficulties in understanding the complex process of replication of bacterial double stranded DNA, investigators have used the single stranded DNA bacteriophage as a test system to understand the exact mechanism of DNA replication.

In this virus, a complementary strand (-) is first synthesised on the parental single strand (+ strand) to yield a double helical structure. The new strand (- strand) then serves as the template for the synthesis of more of + strands which are then incorporated into the viral particles. The process appears to be straight forward but is not as simple as stated. Discontinuous synthesis appears to occur even during this type of replication. Evidence for discontinuous replication of DNA in bacteria first came from Okazaki in 1967, who was able to isolate short pieces of' DNA during replication.

It now appears that not all replicating DNA molecules rely on RNA priming and Okazaki fragments to solve the problem posed by the properties of DNA polymerase. The involvement of t-RNA primers in the synthesis of DNA copies of tumor viruses has also been reported. Recently an explanation to describe the mechanism by which single stranded DNA in certain animal viruses replicates has been given.

It is found that in single stranded DNA, "inverted repeats" (hair pin like structures) at each end of the molecule act as primers for initiation of replication. The secondary structure of the DNA is a double stranded helix. For replication to occur semi conservatively, the structure must undergo unwinding. Over the years, a number of DNA binding proteins, (DNA destabilizing proteins and untwisting enzymes) have been found both in procaryotic and eucaryotic cells and are in-volved in opening up of the double stranded structure to allow replication.

The latest addition to the list are the topoisomerases including the gyrase first isolated from E. coli. This new enzyme catalyses the introduction of negative super coils into double helical DNA in an ATP dependant reaction and appears to be essential for in vivo replication of DNA both in bacteria and bacteriophages. Antibiotics such as nalidixic acid and novobiocin inhibit this enzyme and there by prevent DNA replication at or beyond the replication point.

It is believed to relieve the positive supercoiling strains which builds up during replication and aid in unwinding of the double helix. One other point of interest yet to be solved with regard to DNA replication is the origin of replication and the number of replicating points per DNA molecule. Although a number of origins of replica-tion have been sequenced the mechanism of control is not yet clear. It is however, generally agreed that a DNA molecule may have more than one replicating points from where replication can be initiated,The base sequence at the origin is not identical but has similarities.

A variety of models have been described in literature to explain the mechanism of vivo DNA replication. Unfortunately, none of these provide a complete answer to the problem. The one that has received much attention an d is close to acceptance is the rolling circle model. According to this model, replication starts with a specific cut in one strand of the parental duplex molecule.

This generates a terminal nucleotide with a free 3' OH group while the other end has a phosphate group at the 5' end. As replication proceeds, the 5' end of the open strand is rolled out as a free tail of increasing length. The replicating structure is called a rolling circle since the unraveling of the free single strand is accompanied by a rotation of the double helical template about its axis. The 5' end tail serves as a template for the synthesis of small DNA fragments which are eventually joined together by the DNA ligase. Such growing tails have a double stranded character soon after their formation. Elongation of such tails sometimes goes on to produce tails many times the length of the original circle. It is believed that the tails are cut by specific endonucleases and this is followed by circularization by pairing between the sticky ends.

One of the prevailing fashions in bioscience these days is the application of genomics to eukaryotic gene expression. Largely eclipsed are the approaches of a few decades ago in which enzymes derived from microorganisms blazed the trail to much of our current understanding of macromolecular biosynthesis and gene regulation.

The beginning of the 20th century saw the birth of modern biochemistry with the demonstration that alcoholic fermentation could be observed in the juice of yeast cells. This led to the discovery of the dozen enzymes that convert sucrose to alcohol and ultimately to the reconstitution of alcoholic fermentation at a refined molecular level. Along with these early biochemical studies with yeast came the discovery that virtually the same enzymes and pathway were responsible in mammalian cells for the conversion of glycogen to lactic acid, which provides ATP energy for muscle contraction.

This astonishing fact, along with many other such examples in metabolic and biosynthetic pathways, made it clear that mechanisms and molecules have been preserved in bacteria, fungi, plants, and animals, essentially intact through billions of years of Darwinian evolution. I regard this insight as one of the great revelations of the 20th century. This and other verities gleaned in our lengthy pursuit of the biochemistry of DNA replications can be presented dogmatically as in the THOU SHALTS of the biblical commandments.

RELY ON ENZYMOLOGY

Based on the conviction that all reactions in the cell are

catalyzed and directed by enzymes, the first commandment commands that enzymology can be relied on to clarify a biologic question. Chemists once bridled at this. But time and again, spontaneous reactions, such as the melting of DNA and the folding of proteins, are found to be driven and directed by enzymes; in the case of DNA, its melting in a cell is catalyzed by several different helicases.

The first and crucial step is to find a way to observe the phenomenon of interest in a cell-free system. Should that succeed, then one should be able to reduce the event to its molecular components by enzyme fractionation. This confidence is derived from the fact that, as mentioned, alcoholic fermentation, which had eluded understanding for centuries, was clarified by fractionation of a cell-free yeast extract, as was glycolysis by fractionation of muscle extracts and in the same vein luminescence in the extracts of a firefly and replication of DNA in microbial cell lysates. Fractionation procedures for these extracts revealed the molecular mechanisms and machines for the catalysis and regulation of many complex reactions and pathways, as recounted here for DNA replication.

With a cell-free system in hand that recreates a biologic event, the biochemist should be able to perform the process as well as the cell does it. Even better! After all, the cell is under great constraints to provide a consensus medium that supports thousands of diverse reactions, only some of which operate under optimal conditions. By contrast, the biochemist enjoys the freedom to saturate each enzyme with its substrate, trap the products, and provide the optimal pH and salt and metal ion concentrations. The biochemist can thus be creative and effective in analyzing the molecular basis of a reaction or pathway.

The refinement of methods to purify proteins by chromatography and to establish their homogeneity by gel electrophoresis, when combined with the power of reverse genetics and genomics, has made the isolation and large-scale preparation of enzymes relatively easy compared to what it was years ago. Despite this, discrete events in the proliferation, differentiation, and adaptations of cells and organisms are almost always analyzed by genetic means.

Striking phenotypes are produced by mutations and transfections, but the alterations in enzymes and pathways are generally only inferred. Rarely are they verified by the isolation of proteins with demonstrable functions. To many cell biologists and developmental biologists, the need to examine an event in a cell-free system does not come up on their radar screen.

POWER OF MICROBIOLOGY

The universality of biochemistry from microbes to humans in basic metabolic and biosynthetic pathways has led to the silly quip: "What's true for *E. coli* is true for elephants, and what's not true for *E. coli* is not true." My faith in this universality encouraged me to focus on how prokaryotes, particularly *Escherichia coli*, replicate their own genomes and those of their phages and plasmids. Microbial generation times, unlike those of eukaryotes, are measured in minutes rather than hours and days.

This is where the light on replication shines brightest.we made the choice to work with prokaryotes with the confidence that these systems would be reliable prototypes for how the so-called higher organisms replicate their DNA. In recent years, exciting advances have been made in discovering and characterizing the eukaryotic replication enzymes: the helicases, topoisomerases, polymerases, primases, ligases, and other components of the chromosomal replicases. The variations from prokaryotic enzymes are fascinating. Yet virtually all these enzymes and mechanisms were already familiar and adhere to the basic themes discovered earlier in the prokaryotic systems.

In 1950, having found the enzymes that incorporate nucleotides into coenzymes and curious about how they might become part of nucleic acids, I needed first to determine how the purine and pyrimidine bases became substrates for assembly into these polymers. In the course of exploring the biosynthesis of nucleotides, I learned how to use labeled bases and how to tag each of the phosphates of the nucleoside diphosphates (NDPs) and triphosphates (NTPs).

In 1954, we observed an activity in an extract of *E. coli* that incorporated the label of [á-^{32}P]ATP into an acid-insoluble form that we presumed to be RNA. While making progress in purifying this activity, we learned of a discovery in the laboratory of Severo Ochoa in New York. While observing an exchange of orthophosphate with ADP in extracts of *Azotobacter vinelandii,* they discovered an enzyme that converted the ADP and other NDPs into an RNA-like polymer.

Acting on this information, we substituted ADP for ATP and found that our activity was far greater. Clearly ADP was the preferred substrate over ATP. The *E. coli* enzyme we then purified was the same polynucleotide phosphorylase that the Ochoa laboratory had first identified in *Azotobacter*. As was learned later, the role of the phosphorylase was to degrade RNA rather than effect its synthesis. Had we persisted with ATP as substrate, we would surely have found RNA polymerase, the true synthetic enzyme, a year earlier than we did DNA polymerase and several years before it was discovered in 1961 by the late Sam Weiss.

The late Efraim Racker enunciated this commandment, and I have been one of its ardent disciples. A dramatic example is the discovery of DNA replication. I first observed DNA synthesis in an *E. coli* extract in 1955, when I found that 50 counts out of a million of thymidine were incorporated into an acid-insoluble form. Those few counts above background seemed real because they were susceptible to DNase. Could we possibly figure out what was going on in so crude a system, let alone in an intact cell?

We identified and purified the first DNA polymerase, but not before our fractionation procedures disclosed a variety of novel enzymes that acted initially on the DNA, the [^{14}C]thymidine, and the ATP in our incubation mixtures. First, the [^{14}C]thymidine we added had to be phosphorylated by ATP via a new enzyme, thymidine kinase, to become thymidylate. The added calf thymus DNA proved to be a substrate for DNases that produced the four hitherto unknown deoxynucleoside 52 -monophosphates. These were then phosphorylated by four distinct nucleotide kinases to the corresponding diphosphates, which were in turn phosphorylated by nucleoside diphosphate kinase to the respective and previously unknown dNTPs.

The DNA we had added served three additional functions beyond being a source of the four building blocks. It was a template to direct the precise order of nucleotide assembly, a source of primer termini for chain elongation, and a pool to protect the tiny amount of synthesized DNA from degradation by the nucleases that are abundant in cell extracts.

The *E. coli* extract was thus the source of seven new enzymes in addition to the enzyme we named DNA polymerase. With that, template and primer were introduced into the language of all polymerase actions. We learned all this from fractionating the *E. coli* extract into its multiple activities and finally putting the purified enzymes and their products back together. These studies taught us the basic features of how DNA polymerases act and, incidentally, that the strands of duplex DNA are oriented in opposite directions, not known at the time.

Cell extracts are by their nature "dirty enzymes"; intact cells and organisms are "dirtier" still. F. G. Hopkins, a prescient pioneer in the biochemical basis of nutrition, said it best back in 1931: "(The biochemist's word) may not be the last in the description of life, but without his help the last word will never be said." And so it has been with many cellular events, most recently the fully reconstituted transcription by the 48-subunit yeast RNA polymerase II initiation complex and the sorting of proteins to specific subcellular compartments by fractionated vesicles and enzymes.

Purification of an enzyme to homogeneity now opens the door to reverse genetics, and the enzymes themselves still provide unique reagents, as commandments IX and X will describe. Sometimes, an apparently pure enzyme may be found on further purification to harbor a contaminant of great importance.

As one example, an extra step in the purification of DNA

polymerase I rendered the enzyme inactive. The reason: the template-primer used in our routine assays was DNA activated by being nicked many times by a DNase. We were not aware at the time that these nicks had been enlarged upon by an exonuclease in our polymerase preparation to create a stretch of exposed template needed by the polymerase. The exonuclease activity we had fractionated away, which we then purified and named exonuclease III, proved to be a crucial reagent in the discovery of recombinant DNA. The importance of purifying an enzyme was such that a student of mine had her personalized California license plate read: PURIFY.

In 1967, after 10 years of trying and failing to prove that our enzymatically synthesized DNA was biologically active, we finally succeeded. What made the difference was the use of a single-stranded, circular DNA of a bacterial virus as template and the discovery of DNA ligase that could circularize the linear product. With X174 DNA, we could make quantities of the circularly closed, infectious viral DNA. The sequence of 5,386 nucleotides was correct, and there was no need for novel nucleotides or other components, as had been conjectured. We also pointed out that this in vitro system afforded the means to introduce novel nucleotides for site-directed mutagenesis.

The appearance of our paper announcing the test tube synthesis of infectious DNA generated a huge crush of media attention, a congratulatory phone call from President Lyndon Johnson and headlines worldwide, all based on the belief that we had synthesized a big, hairy virus and "created life in the test tube." I had to explain to the assembled reporters that it was not I who assembled the long DNA chain of the virus but rather it was the awesome enzyme, DNA polymerase, that I had identified and isolated from *E. coli* cells. And further, I had to make it clear that it was these bacteria in a culture flask that imbibed the viral DNA to make the infectious virus particles. As for "creation of life in the test tube," some might dispute that a virus is even a "living" creature.

Three years after the hoopla about the synthesis of a viral DNA by our DNA polymerase, serious questions remained.

How is a DNA chain started? How is the accumulating genetic evidence for additional polymerases and other factors needed for replication explained? A cartoon at the time showed the apparatus at a replication fork discreetly obscured by a fig leaf, and there were polemical attacks in *Nature New Biology* that dismissed our DNA polymerase as merely a repair enzyme with little relevance to replication, in essence a "red herring."

Over the years, we had tried a variety of DNA samples to demonstrate the start of a DNA chain. The results were negative or equivocal. Then it dawned on me that we were violating the fifth commandment. We were using a pure DNA polymerase on a dirty DNA substrate: frayed, gapped, fragmented, denatured and heterogeneous. When we finally switched to the intact, single-stranded, circular DNA of a small bacteriophage, we discovered how a DNA chain is started: priming with RNA.

Single-stranded phages provided not only the DNA substrate with which we could discover the RNA priming of new chains, but also the enzyme systems responsible for the priming and subsequent replication.

The filamentous phage M13 depends on the host RNA polymerase to make a short transcript of an origin region, whereas the icosahedral phage X174 appropriates a complex primosome, used by the host to prime the start of chains on the lagging strand at the replication fork. Whereas the conversion of the single M13 viral strand to the duplex replicative form was readily resolved and reconstituted, the conversion of the X174 single-stranded circle was far more complex and required the discovery of 15 new proteins, which constitute the apparatus at the host chromosomal replicating fork. With these many proteins in hand we could attempt to discover how replication was initiated at the origin of the intact *E. coli* chromosome.

MOLECULAR CROWDING

Cells are gels. Half of the cell dry weight is made up of proteins packed in highly organized communities. That some of their functions, individually and collectively, can be observed despite great dilution (20-fold or more) is a fortunate break for biochemistry. But there is an absolute need in some cases to restore the crowded molecular state, as we learned from our attempts to observe initiation of replication at the origin of an intact chromosome.

We were given a 5-kb plasmid containing the origin of the 4,000-kb *E. coli* chromosome that is replicated in the cell with the physiological and genetic features of the host chromosome, in effect a minichromosome. When Seichi Yasuda came from Japan with this *ori*C plasmid, I thought we would soon resolve and reconstitute its replication much as we had done with phage X174. But it took 10 man-years of utter frustration before we finally succeeded in making a cell-free system work.

Success in achieving *ori*C plasmid replication in a cell-free state depended on two strange maneuvers. One was to include a high concentration of polyethylene glycol (PEG) (10% [wt/vol]), 10,000 Da) in the incubation mixture. As is true of such hydrophilic polymers, the PEG gel occupies most of the aqueous volume and excludes a small volume into which large molecules are crowded. This concentration is essential when several proteins are needed in the consecutive steps of a pathway. The other maneuver repeated an earlier experience in which we proceeded to fractionate an inactive lysate with ammonium sulfate. Progressive additions of the salt yielded precipitates, in one of which the active proteins were present and concentrated when dissolved in a small volume. Just as important, the supernatant fraction we discarded contained a potent inhibitor, a nuclease that relaxed the plasmid DNA from its essential supercoiled state.

Along with a purified, origin-binding DnaA protein, we provided primosomal, replication, and ligase proteins with other factors to obtain rapid, origin-specific, extensive replication of the *ori*C minichromosome. With these many proteins added in sufficient amounts, PEG was no longer needed. The mechanisms we discovered in *ori*C replication were found to apply to the replication of many microbial plasmids and phages and to some eukaryotic viruses and episomes.

RESPECT THE PERSONALITY OF DNA

For years, DNA was regarded as a rigid rod devoid of personality and plasticity. Only upon heating did DNA change shape, melting into a random coil of its single strands. Then we came to realise that the shape of DNA is dynamic in ways essential for its multiple functions. Chromosome organization, replication, transcription, recombination, and repair have revealed that DNA can bend, twist, and writhe, can be knotted, catenated, and supercoiled (positive and negative), can be in A, B, and Z helical forms, and can breathe.

Especially noteworthy is breathing, the transient thermodynamic-driven opening (melting) of the duplex that facilitates the binding of specific proteins such as the helicase responsible for priming and the onset of replication. Certain DNA sequences are also predisposed to a more extensive form of melting ("heavy breathing") that creates a relatively large opening for transcription.

The resulting RNA-DNA duplex (R-loop) can activate an inert origin of replication by altering its structure, even hundreds of base pairs away, which facilitates its opening by origin-binding and replication proteins. Negative supercoiling supplies the energy for the breathing and other features that direct the shape and movements of DNA at the *oriC* origin of replication. These DNA responses have led to an appreciation of the role of transcriptional activation of replication origins near primers in large chromosomes, both prokaryotic and eukaryotic.

USE REVERSE GENETICS AND GENOMICS

Direct genetics, in which a randomly mutated gene can ultimately be linked to a deficiency in a single enzyme, was a landmark discovery in biologic science. This approach served well by providing *E. coli* mutants defective in replication, some in initiation of a chromosome (e.g., *dnaA*) and others in elongation (e.g., *dnaB, dnaC, dnaE,* and *dnaG*). But randomly generated mutants do not readily disclose the products of their genes nor their particular functions. Nevertheless, these replication mutants were crucial in validating our assays because DNA synthesis was absent in the extracts of mutant cells and restored when extracts or purified fractions from wild-type cells were added.

Reverse genetics and genomics have now made the enzymologic approach even more powerful. Unlike direct genetics, enzymology starts with a defined function, after which finding the responsible genes has become relatively easy. With even a picomole of a purified enzyme or a band on a gel, a peptide sequence can be determined and the encoding gene identified, cloned, and overexpressed; genomics facilitates the process by providing the complete genome sequences of *E. coli*, yeast, and many other microbes. Profound insights into the physiologic role of an enzyme or pathway emerge from the behaviour of cells with a null, point, or truncated mutation of a gene or modulated levels of its overexpression.

The ease with which large quantities of pure enzymes can be produced by overexpression of a cloned gene has made their use as reagents even more attractive (commandment X). "DNA shuffling", a new technique, has made enzyme reagents compelling. By creating a very large number of random rearrangements of a gene or genome, a particular gene product can be selected for a desired property (e.g., heat resistance) with wide applications in industry and biomedical science.

EMPLOY ENZYMES AS UNIQUE REAGENTS

Biochemistry is replete with examples in which enzymes have been employed as analytic and preparative reagents. From basic research to industrial processes, proteases, amylases, phospholipases, kinases, and phosphatases, etc., have been crucial in operations that were beyond the capabilities of available chemical technology. I will mention just a few examples of applications to DNA and its replication and one from my recent research on inorganic polyphosphate (poly P).

The key discovery in 1944 that identified DNA as the genetic substance was based on the destruction by crystalline pancreatic DNase of the factor that transformed one strain of *Pneumococcus* sp. to another. It was the action of this DNase again, which in 1955 made me believe that the few counts of

[^{14}C]thymidine incorporated by an *E. coli* extract into an acid-insoluble form signaled the synthesis of DNA. Many more examples can be cited in which an enzyme reagent was decisive: the circularization of linear DNAs by ligases, the creation of "sticky" tails by specific exonucleases used to prepare the first recombinant DNAs, the innumerable uses of restriction nucleases, and on and on.

Enzyme reagents have been decisive in my approach to determine the functions of poly P, an inorganic polymer of hundreds of phosphate residues linked by "high-energy" anhydride bonds. Likely present on prebiotic earth, poly P is now found in every living cell, but for lack of any known functions, was earlier regarded as a "molecular fossil." True to the first commandment, I have sought and isolated enzymes that make and act upon poly P.

With these enzymes we developed assays that are definitive, facile, and sensitive in place of those that are ambiguous, laborious, and insensitive. Together with the use of reverse genetics, we have learned that many microbes need poly P to adapt to adverse conditions and to survive in the stationary phase. The kinase that makes poly P from ATP is highly conserved in some of the major pathogenic bacteria, and mutants lacking the kinase are defective in motility and virulence. Thus, this enzyme, absent from eukaryotes, may prove to be an attractive target for antimicrobial drugs.

INDUCED DNA DAMAGE

Two main pathways appeared during evolution: the release of oxygen by photosynthesis and the aerobic respiration. These highly efficient metabolic systems produce useful, although dangerous molecules for the cells: free radicals (NO· ¼) and reactive oxygen species (ROS) such as $O_2 \cdot^-$, H_2O_2 and OH·. These molecules are constantly formed in cells by the cellular metabolism and by spontaneous chemical degradation of some biomolecules. Free radicals are implicated in a wide range of cellular processes but an excess of such molecules could have deleterious effects on living cells leading to cell injury or cell death. Oxidative damage of main cellular components such as DNA, lipids, carbohydrates and proteins has been implemented in the ageing phenomena, ischemia, cancer, autoimmune diseases and neural cell death.

Activation of oncogenes like c-*myc* and ras can also induce ROS by alteration of specific metabolic pathways. Organisms are equally exposed to exogenous factors such as ionising radiation and chemical cancerogens, which can also generate ROS. Evidence has accumulated that lack of protection against free radicals and lack of repair of oxidative damage in biological macromolecules have a significant role in mutagenesis and consequently on carcinogenesis.

Thus maintenance of the cellular equilibrium between prooxidant (ROS) and antioxidant species should be tightly regulated in cells. Specialised systems have evolved, involving cellular antioxidants and DNA repair, to protect cells against ROS-induced injury. Indeed, the defence system, such as DNA repair, is highly conserved from bacteria to human.

It is generally assumed that most oxidative DNA damages - base damage, sugar damage and abasic sites - are dealt with by BER. A number of recent reviews report the main features of the BER pathway. The goal of DNA glycosylases is to locate fast and efficiently the aberrant base amongst a huge excess of normal ones. Very little is known about how these proteins achieve this goal. The comparison of the crystal structures of a number of DNA glycosylases revealed structural homologies leading to the concept of a superfamily of BER DNA glycosylases, the helix-hairpin-helix (HhH) superfamily, having similar HhH fold and a Gly/Pro-rich stretch with nearby Asp (GPD) motifs, although very little sequence similarity.

This HhH motif plays an important role in the flipping out of the modified base. The rate of repair measured for the excision of modified bases is not always optimal and should be improved by the identification and the use of accessory proteins. The recent identification of new DNA polymerases able to replicate efficiently and accurately miscoding and modified bases have to be taken into account in the understanding of BER.

OXIDATIVE DAMAGE OF DNA

DNA has a limited chemical stability (intrinsic or induced by exogenous agents) and is one of the most biologically important targets of ROS. Maintenance of its integrity is a major goal for cells. About 100 different kinds of base and sugar damage have been identified. Free-radicals can damage nucleobases and sugar units in DNA either directly, or indirectly. Hydroxyl radicals, which are the most active species, predominantly, react with the C_8 of purines forming 7,8-dihydro-8-oxo-2'-deoxyguanosine (8-oxoG) and imidazol ring-opened products such as 2,6-diamino-4-hydroxy-5-formamidopyrimidine (Fapy); with C_5-C_6 double bond of pyrimidines forming glycol and pyrimidines hydrates and with C8-C5' of purines forming 8,5'-cyclopurine deoxynucleosides.

The abstraction of a hydrogen atom from deoxyribose at C1' and C4' generates DNA-strand breaks with 3'-phosphoglycolate ester and 3'-phosphate. Indirectly, ROS can generate reactive aldehydes, as a product of membrane lipids peroxidation, which react with DNA bases forming the exocyclic adducts 1,N^6-ethenoadenine, 1,N^2-ethenoguanine, N^2,3-etheno-guanine, and 3,N^4-ethenocytosine and pyrimidopurinone such as M_1G.

Damaged bases such as 8-oxoguanine; 5-hydroxy-2'-deoxycitidine, hypoxanthine, ethenoadducts and pyrimidopurinone have miscoding properties and, if not repaired, lead to mutation upon replication. Others such as oxidised deoxyribose, formamidopyrimidine, fragmented thymine and thymine glycol cause replication block and therefore are believed to have a strong cytotoxic effect.

Ionising radiation induces mutation and chromosomal aberrations in cells, which are believed to lead to cancer and loss of neural function in humans. Humans are daily exposed to low doses of radiation during air travel or from radon in homes and areas of low-level contamination. Energy from ionising radiation, such as X-rays and gamma rays, is transmitted in the water surrounding the DNA molecule in such a way that between 2 to 5 radical pairs are generated within a radius of 1 to 4 nm. Clustered multiple damaged sites induced by ionising radiation, have been observed, most of them being modified bases rather than DNA strand breaks. The complexity of radiation-induced clustered DNA-damage depends on the ionising density of LET (Linear Energy Transfer) radiation.

Ionising radiation was hypothesised to produce clustered damage, and clusters spanning several kilobase pairs to a few base pairs were modeled. Irradiation of DNA in non-radioquenching solution with gamma rays was shown to induce 1 double strand break to ~0.5 Nth protein-recognised oxidised pyrimidine cluster: 1.5 Nfo protein-recognised abasic cluster: 2 Fpg protein-recognised oxidised purine clusters. Use of the polyamines putrescine to measure abasic clusters refractory to Nfo protein cleavage reveals twice as many abasic clusters in -irradiated DNA as does Nfo, suggesting that even more clusters are produced than these values indicate.

The dose-response relations in a log-log plot are straight lines with slopes near 1, showing that clusters are formed by a single radiation hit. All DNA lesions are not equally frequent components of clusters, with about 15% of oxidised pyrimidines, oxidised purines and abasic sites in clusters, but only about 8% of the strand breaks are in DSBs.

Since double strand break yields are ~100 times higher in the absence of radical scavengers fewer clusters are expected in DNA irradiated in the presence of a radical scavenger such as Tris. Indeed the absolute levels of bistranded cluster levels are significantly reduced in radioscavenging solution, and, surprisingly, the ratios of specific cluster types also changed strikingly. Cluster levels in human cells exposed to 50 kVp X-rays shows that X-rays induce 1 DSB: 0.75 Nfo-abasic cluster: 1 Fpg-oxidized purine: 0.9 Nth-oxidized pyrimidine cluster. In cells, non-DSB clusters are at least ~70% of the complex damages.

Clusters are postulated to be critical because they may be more difficult to repair than random lesions. In contrast to randomly located oxidised bases, clustered modified bases are within half a turn of the double helix, i.e. 5 nucleotides and some of them on the two strands, therefore the excision repair pathways would have difficulty in removing clustered lesions. It is anticipated that the excision/incision of oxidatively damaged bases, including AP sites, on both strands will, if not tightly regulated, either inhibit certain steps of repair or produce double strand breaks and thus be lethal for the cells. Studies of purified enzymes acting on oligonucleotides with defined lesions at specific relative spacing on opposing strands indicate that clusters may comprise non-repairable, highly repair-resistant and pre-mutagenic damage, and that lesion spacing and polarity are important. However, the precise repair mechanisms for the clustered lesions are so far very poorly understood.

E. coli generate high levels of DSBs during repair after irradiation, and the repair-generated DSBs likely result from abortive cluster repair. Although rodent cells exposed to high doses (500 Gy) also increase DSB levels, their origin and whether they are generated at low doses was not clear.

BASE EXCISION REPAIR

Oxygen radicals generate mostly non-bulky DNA lesions, most of them are substrates for BER enzymes. Only few oxidative DNA lesions such as cyclopurine and pyrimidopurinone are substrates for nucleotide excision repair (NER) enzymes. Key enzymes of the BER pathway are DNA-glycosylases. They remove damaged and mispaired bases from DNA by cleavage of the *N*-glycosylic bond between the abnormal base and deoxyribose, leaving an abasic site in DNA. Most DNA glycosylases are highly specific and can excise various types of modified bases.

There are two types of DNA glycosylases: mono- and bifunctional. The mono-functional ones cleave the *N*-glycosylic bond releasing the modified base and generating an AP site as a final product which in turn is recognised by an AP endonuclease. The bifunctional ones cleave the *N*-glycosylic bond, liberate the modified base and in a concerted manner cleave the phosphodiester bond 3' to the resulting AP site by a or - elimination mechanism (-lyase activity) generating a single-strand break with 3'-phosphate and 3'-phosphoglycolate (PGA) extremities respectively. Then the DNA backbone next to the abasic site or 3'-phosphate/PGA terminus is cleaved by an AP-endonuclease allowing a DNA polymerase to fill the gap before DNA ligase reseals the DNA. Experiments with *E. coli* mutants deficient in BER pathway demonstrated the crucial role of DNA glycosylases and AP-endonucleases in protecting cells from mutagenic and cytotoxic effects of free radicals.

In human cells BER proceeds via two alternative pathways, either 'short-patch' DNA polymerase -dependent pathway which involves the replacement of a single nucleotide or 'long patch' PCNA-dependent pathway which involves the replacement of up to six nucleotides. The former pathway was reconstituted *in vitro* with the purified human proteins uracil-DNA glycosylase (UDG), AP endonuclease 1 (APE1), DNA polymerase (pol), the scaffold protein XRCC1 and DNA ligase (I or III). Long-patch pathway can be achieved with AP endonuclease, Pol and PCNA.

The antibodies directed against PCNA totally suppress repair patches longer than one nucleotide. Reconstitution *in vitro* of long-patch repair showed that DNase IV/FEN 1 is required in addition to proteins implicated in short-patch pathway and that PCNA greatly stimulates the reaction. Either DNA polymerase/ or DNA polymerase have been shown to perform the synthesis step in this sub-pathway. Recently, it has been demonstrated that removal of 8-oxoG could be achieved when using only hOGG1, Ape1, pol and DNA ligase I.

BASE EXCISION REPAIR IN PROKARYOTES

Mechanistic studies of both classes of DNA glycosylases led to formulation of a model of unified catalytic mechanism. Monofunctional DNA glycosylases cleave the glycosidic bond by a hydrolytic mechanism, activating a water molecule to attack the C1' of the damaged base, resulting in AP site as a final product. In contrast, the DNA glycosylases/AP lyases employ a nucleophilic group in the enzyme to perform attack at C1'. In all DNA glycosylases/AP lyases so far characterised, an amino group has been implicated as a nucleophile. As a result of nucleophilic attack by amino group, a covalent imino intermediate (Schiff base) is formed between the C1' of the lesion and the enzyme.

By abstracting the C2'-*pro-S* proton, the enzyme initiates electron rearrangement that leads to the release of the 3'-phosphate. The Schiff base is then hydrolysed, releasing the enzyme and leaving a nick with an,-unsaturated aldehyde at the 3'-end and a phosphate at the 5'-end of the DNA. Upon completion of -elimination, some enzymes such as the Fpg protein abstract proton at the C4' and promote a -elimination. The resulting structure is a single-base gap with phosphates at both 3'- and 5'-ends.

Sodium borohydride and cyanoborohydride have been used to show that the reaction involves a Schiff base intermediate. When the reduction reaction occurs in the glycosylase-DNA complex, the enzyme becomes irreversibly crosslinked to DNA.

The Fpg protein (formamidopyrimidine-DNA glycosylase): Among the DNA glycosylases, the *E. coli* Fpg protein (formamidopyrimidine-DNA glycosylase/MutM protein) has been one of the most extensively studied. The gene coding for the Fpg protein was cloned and the physical and enzymatic properties of the protein established. It is a globular monomer of 30.2 kDa of 269 amino acids. *In vitro*, the Fpg protein excises a broad spectrum of modified purines, particularly 2,6-diamino - 4 - hydroxy - 5N - methylformamidopyrimidine (Fapy) and 7,8-dihydro-8-oxoguanine (8-oxoG) residues. Moreover, the Fpg protein is also able to excise various pyrimidine oxidation products such as 5-hydroxycytosine and 5-hydroxyuracil, the ring fragmentation product of thymine (RT), thymine glycol and 5,6-dihydrothymine.

The Fpg protein is a bifunctional DNA glycosylase, endowed of an AP lyase activity that incises DNA at abasic sites by a --elimination mechanism and an activity excising 5'-terminal deoxyribose phosphate (dRPase). *In vivo*, the Fpg protein has an antimutator effect preventing G/CT/A spontaneous transversion. It acts in concert with the MutY protein. The *fpg mutY* double mutant CC104 has an extreme mutator phenotype that can be reversed by plasmids carrying the fpg gene. Interestingly, expression of the bacterial fpg gene in mammalian cells reduces the mutagenicity of -rays but has no effect on survival.

The Fpg protein in its COOH-terminus contains the zinc-finger motif (Cys-X_2-Cys-X_{16}-Cys-X_2-Cys-X_2-COOH) that is mandatory for Fpg binding to DNA and to its enzymatic activities. The active site of Fpg protein is located within the first 73 amino acid residues of the amino terminus. The targeted mutagenesis of conserved amino acid residues was used to elucidate the mechanism of enzymatic catalysis. It was found that conserved residues lysine 57 (K57G), lysine 155, glutamates 2, 5, 131 and 173 and proline 2 (P2G) dramatically reduce the cleavage of oxidised bases such as 8-oxoG. Recently, major progress has been achieved in solving the three-dimensional structure of the Fpg protein from various origins.

Two Fpg proteins have been isolated from the highly radioresistant bacteria *Deinococcus radiodurans*. Both excise Fapy and 8-oxoG residues and present a -lyase activity. In addition one excises thymine glycols. The genes have been cloned and the detailed specificities established using pure proteins.

The Nth protein (Endonuclease III): The Nth protein was initially identified as an endonuclease that specifically cleaves DNA damaged by X-rays, UV light and free radicals. It is a DNA glycosylase with a broad substrate specificity, excising ring-saturated, ring-opened, and ring-fragmented pyrimidines, such as thymine glycol, 5,6-dihydrothymine, 5-hydroxy-6-hydrothymine,5,6-dihydrouracil, alloxan, 5-hydroxy-6-hydrouracil, uracil glycol, 5-hydroxy-2'-deoxy-cytidine, 5-hydroxy-2′deoxyridine, -ureidoisobutiric acid and also -R-hydroxy--ureidoisobutiric acid, a fragmentation product of the 5R-thymidine C5-hydrate.

Unexpectedly, it has been demonstrated that Nth is implicated in the repair of 8-oxoG, since it removes 8-oxoG from 8-oxoG/G mispairs and the triple mutant *fpg nth nei* displays increased spontaneous G/CC/G transversions as compared to single and double mutants in these three genes. This leads to the proposal of a new model in which Nth and Nei repairs 8-oxoG in nascent and transcriptionally active DNA. The nicking activity at abasic sites of Nth is due to its AP lyase function. The phosphodiester bond cleavage occurs via -elimination (the protein does not cleave at reduced AP sites) generating 5' ends bearing 5'-phosphate and 3' ends with 2,3-unsaturated abasic residue 4-hydroxy-2-pentanal. This terminus requires further processing to be used for DNA repair synthesis.

The gene encoding for the Nth protein, the *nth* gene, was cloned and protein purified to homogeneity. The three dimensional structure of Nth was also established. The Nth protein is a monomeric protein of 23.4 kDa (211 aa). It is an elongated protein with a cleft separating two similar size domains: a continuous domain formed by six--helices and a second domain formed by three C-terminal -helices and the N-terminal helix. The C terminal loop contains an iron-sulphur centre (4Fe-4S) that is anchored to the protein by a Cys-X6-Cys-X2-Cys-X5-Cys sequence. This cluster has a structural function in positioning basic residues for DNA-binding and it is also present in the MutY protein.

E. coli nth mutants deficient in the Nth protein does not show apparent phenotype. They are not sensitive to X-rays, H_2O_2 or other agents that produce ring saturation and fragmentation products of pyrimidines in DNA. However the double mutant *nth nei* exhibits a strong spontaneous mutator phenotype and it is hypersensitive to ionising radiation and H_2O_2.

The Nei protein (endonuclease VIII): The capacity of *E. coli nth* mutants defective in most of the activity to excise thymine glycol, allowed the identification of a new activity excising this oxidised base, that was named endonuclease VIII or Nei protein. It has been purified from *E. coli* extract lacking the Nth protein. Similarly to the Nth protein, Nei removes the oxidised forms of thymine such as thymine glycol, dihydrothymine, -ureidoisobutyric acid, and urea residues when present in DNA and cleaves phosphodiester bond at

AP sites. The cloned *nei* gene encodes a 263 amino acid polypeptide with a striking similarity to the Fpg protein.

The catalytic site's residues are highly conserved in these two proteins, however, their substrate specificities are different. Characterisation of apparently homogenous Nei protein has shown that it can equally remove the oxidised forms of cytosine such as 5-hydroxycytosine and 5-hydroxyuracil. More recently, genetic evidence for Nei implication in the prevention of spontaneous GT transversions was confirmed since Nei was found to cleave 8-oxoG efficiently when opposite to A and G.

Furthermore, extended the substrate specificity of the Nei protein and have demonstrated that Nei and Nth significantly differ from each other in terms of kinetic although they share common substrate specificity. The crystal structure at 1,25 Å of the Nei covalent intermediate complex with DNA largely confirms that its structure is similar to that of Fpg.

MONOFUNCTIONAL DNA GLYCOSYLASES

AlkA protein: The main substrates for the *E. coli* AlkA and Tag I proteins are alkylated bases. The Tag I protein is a major DNA glycosylase removing 3-methylpurines residues. However, the AlkA protein in contrast to Tag I is also involved in repair of oxidative DNA damage. Hypoxanthine (HX) is generated in DNA by spontaneous deamination of adenine and also by the free radical nitric oxide. HX residues in DNA are mutagenic since they can pair with cytosine, generating AT to GC transitions after DNA replication. Hypoxantine-DNA glycosylase releases HX from DNA containing dIMP.

Cloning the gene coding for the HX-DNA glycosylase in *E. coli,* it was shown that the 3-methyladenine-DNA glycosylase coded by the alkA gene carries the HX-DNA glycosylase activity. ANPG, APDG and MAG proteins, the human, rat, and yeast functional homologues respectively of AlkA protein excise HX residues when present in DNA, the mammalian enzymes being most efficient. The AlkA protein also catalyses the excision of ethenobases N^2,3-ethenoguanine and 1,N^6-ethenoadenine.

The 5-formyluracil (5-foU) residue, an oxidised thymine lesion is a major DNA damage induced by ionising radiation. It induces A/T to G/C transition and is repaired in *E. coli* by the AlkA protein. In addition, AlkA protein excise RT residues when present in DNA with an apparent K_m=170 nM. Thus the AlkA protein has very broad substrate specificity. The three dimensional structure of AlkA reveals a compact globular protein with a prominent hydrophobic cleft on its surface and three equal-sized domains.

Uracil DNA glycosylase (UDGs) super-family: Base excision repair was first established for uracil. Four sub-families of uracil-DNA glycosylase (UDG) have been identified. The best studied family of UDGs is *E. coli* Ung protein. Ung is specific for uracil and it is present in a wide range of living organisms from bacteria, lower to higher eukaryotes, DNA viruses and human. So far UDG has not been implicated in the repair of oxidative DNA damage. The second family corresponds to the mismatch-specific uracil-DNA glycosylase (MUG) identified in some eukaryotes and in several prokaryotes. MUG excises thymine from G/T mismatches in DNA and it is also active on mispaired uracil in G/U pair. The crystal structure of Ung and Mug proteins show that they are structurally similar, despite low sequence homology.

Another sub-family has been characterised from thermophilic archea and several bacteria. These double strand UDGs (dsUDG/DUG) are able to remove uracil from double strand DNA either in U/A and U/G context. The fourth sub-family, initially identified in vertebrates is represented by the human single-strand mismatch-specific uracil-DNA glycosylase (hSMUG1). However hSMUG1 is more active on double-stranded than on single-stranded DNA, it prefers uracil opposite A and G. The 5-hydroxymethyluracil residue is a product of oxidation and subsequent deamination of 5-methylcytosine. Teebor's group identified the mammalian 5-hydroxymethyluracil DNA N-glycosylase activity as SMUG1. A fifth sub-family has been characterised recently and is represented by *pa*-UDGb isolated from *Pyrobaculum aerophilum*. *pa*-UDGb has broad substrate specificity. It can remove uracil and also hypoxanthine when present in DNA. Among the members of the UDG family several enzymes are implicated in the repair of exocyclic DNA adducts which are highly mutagenic oxidative DNA lesions. The *E. coli* MUG protein is able to remove 3,N^4-ethenocytosine, 8-(hydroxymethyl)-3,N^4-ethenocytosine and 1,N^2-etheno-guanine when present in DNA. *MutY:* The *E. coli mutY* gene encodes a DNA glycosylase of 39 kDa, which excise A opposite G and C. MutY also removes A opposite to 8-oxoG and 7,8-dihydro-8-oxoadenine (8-oxoA).

The protein shows significant sequence homology to Nth and is also an iron-sulphur protein which contains the conserved (4Fe-4S) cluster/HhH domain. The protein was extensively characterised using various approaches including site-directed mutagenesis. However, there is still controversy regarding the AP lyase activity of the MutY protein. Proteolytic cleavage of MutY generates two fragments of 26 kDa and 13 kDa. The 26 kDa fragment corresponds to the catalytic core domain, it retains normal DNA binding and adenine-DNA glycosylase activity but lacks its activity against A/8-oxoG.

The specificity for 8-oxoG residues is due to the C-terminal domain of MutY. It is thought that MutY plays an important role in the maintenance of genome integrity following oxidative DNA damage. The presence of AP endonucleases (Xth and Nfo) greatly enhances the excision rate of A when opposite to G by MutY.

These data suggest that the MutY-DNA complex interacts with Nfo and Xth proteins *in vivo*. The crystal structure of MutY catalytic core domain (cMutY) has been solved. cMutY is an all- protein that retains the canonical bilobal architecture of Endonuclease III. The compression of intra-strand phosphate distance by HhH and pseudo HhH domains permit the DNA bending and the nucleotide flipping.

The flipped nucleotide is recognised by specified residues of the active side pocket located between the two domains. Crystal structures and mutagenesis results define that MutY cleaves the N-glycosylic bond through a hydrolytic mechanism requiring Asp138, with uncoupled, inefficient AP lyase activity due to Lys142.

AP ENDONUCLEASES: XTH AND NFO

Abasic sites (AP sites) occur in DNA through spontaneous depurination/depyrimidination, by the action of ROS and as secondary lesions after excision of modified bases by DNA glycosylases. AP sites have miscoding properties since the replication machinery incorporates preferentially adenine opposite to AP site. This observation was later named 'A rule'.

The occurrence of an enzyme recognising AP sites in *E. coli* was first described in the seminal work of Verly. Two types of enzymes recognise and excise DNA at AP sites: AP endonucleases such as Nfo and Xth proteins act on this lesion by a hydrolytic mechanism whereas enzymes carrying a -lyase activity such as Fpg, Nei and Nth proteins incise the AP site via a - and elimination mechanism respectively.

THE NFO PROTEIN (ENDONUCLEASE IV)

The Nfo protein is an EDTA-resistant AP endonuclease that represents only 5% of the total AP endonucleasse activity in *E. coli* wild type. It is inducible by oxidative stress and is under the control of the soxRS system. The gene coding in *E. coli* for Endonuclease IV, *nfo*, has been cloned, and the protein characterised. It is a monomer of 30 kDa having several activities: AP endonuclease, 3'-phosphatase and 3-phosphoglycoaldehyde diesterase activities. It has been suggested that these 3' repair activities clean the ends, a prerequisite for DNA synthesis. *E. coli nfo* mutants deficient in the product of the *nfo* gene are extremely sensitive to the lethal effects of oxidative agents such as bleomycin and t-butyl-hydroperoxide. In addition, this mutation enhances the sensitivity of *xth* mutants to H_2O_2, and alkylating agents.

The double mutants *xth nfo* are also sensitive to -radiation. We will discuss below the role of the Nfo protein and its yeast homologue Apn1 in the nucleotide incision repair pathway (NIR). The high-resolution structure of Nfo protein has been solved. It shows that Nfo is an protein arranged as a $_{88}$-barrel. This structure, well suited for the binding to large molecules such as DNA, contains three Zn^{2+} ions at the active site which are critical for the activity. The triple Zn centre is ligated to protein by conserved residues that cluster at the centre of the crescent-shaped deep pocket. The Nfo protein detects AP site by insertion of its side chains into the DNA minor groove, it flips the target AP site and the opposite nucleotide out of the DNA base stack to produce a 90° bend in the DNA.

The Xth Protein (Exonuclease III)

The Xth protein, coded for by the xth gene, originally identified as a 3'5' exonuclease active on double-stranded DNA and a 3'-phosphatase, is the major AP endonuclease of *E. coli*: over 80% of the total AP endonuclease activity in wild type. This protein is also active on single stranded DNA. Beside its exonuclease and 3'-phosphatase activities it has 3' repair diesterase and ribonuclease H activities. The 3' termini generated by Xth are normal nucleotides with 3' hydroxyl group that are effective primers for DNA polymerases.

The *xth* mutants are extremely sensitive to H_2O_2, and to oxidative DNA damage generated by near-UV light. The crystal structure of the Xth protein has been solved, and it has been assumed that the base opposite to the abasic site plays an important role for the recognition of the lesion.

Mechanism of action and three-dimensional structure of bifunctional DNA glycosylases: the Fpg case: It was hypothesised that DNA glycosylase/AP lyases use a mechanism involving the nucleophilic attack on the C1' prime of the modified deoxynucleoside targeted for excision, thus displacing the aberrant base and forming a transient intermediate (Schiff base) with the C1'-deoxyribose moiety. The amino acid sequence of the Fpg protein begins with a Met-1 which is processed, thus Pro-2 being the N-terminal.

The N-terminal proline (P2) residue of the Fpg protein, a highly conserved residue, was shown to be linked with DNA containing 8-oxoG residues, suggesting that proline is the nucleophile initiating the excision of 8-oxoG. Site-directed mutagenesis demonstrated the mandatory role of the N-terminal proline residue in the 8-oxoG-DNA glycosylase activity of the Fpg protein *in vitro* and *in vivo*, as well as in its

AP lyase activity upon pre-formed AP sites but less of a role in Fapy-DNA glycosylase activity.

The role of the conserved lysine 57 and 155 (K57 and K155) residues upon the various catalytic activities of the Fpg protein was examined by targeted mutagenesis. The lysine 57glycine (FpgK57G) mutant protein had a dramatically reduced (55-fold) DNA glycosylase activity for the excision of 8-oxoG residues. The FpgK57G protein was poorly effective in the formation of Schiff base complex with 8-oxoG/C DNA. However, the mutant could partially restore the ability to prevent spontaneously induced G/CT/A transversions in *E. coli* BH990 (*fpg, mutY*) cells. The DNA glycosylase activity of FpgK57G using FapyGua residues as the substrate was comparable to that of the wild type enzyme. These results suggested that K57 could participate either in a direct interaction with the C_8 oxygen of the 8-oxo purines or in the opening of the furanose ring, in the first step of the model proposed for the mechanism of action of the Nth protein.

Effect of mutations of conserved glutamic and aspartic acid residues to glutamines and asparagines, have been studied. While the Asp to Asn mutants had no effect on the incision activity on 8-oxoG DNA, several of the substitutions at glutamates reduced Fpg activity on the 8-oxoguanosine DNA, with the E3Q and E174Q mutants being essentially devoid of activity. Unexpectedly, the AP lyase activity of all of the glutamic acid mutants was slightly reduced as compared to the wild-type enzyme. Furthermore, it has been shown that lysine 57 but not proline 2 is crucial for catalysis of oxidatively damaged pyrimidines.

Sugahara and co-workers determined the crystal structure of Fpg homologous protein from an extreme thermophile, *Thermus thermophilus* HB8 (HB8-Fpg) at 1.9 Å resolution. It reveals that the Fpg protein molecule is composed of two distinct domains connected by a flexible hinge. More recently, Castaing and colleagues resolved the structure of a non-covalent complex between the *Lactococcus lactis* Fpg and a 1,3-propanediol (Pr) abasic site analogue-containing DNA. They have shown that Fpg pushes out the Pr site from the DNA double helix, recognising the cytosine opposite the lesion and inducing a 60 degree bend of the DNA. Finally, the structure of a trapped catalytic intermediate of *E. coli* Fpg protein has been determined at 2.1 Å resolution. In agreement with the structure of HB8-Fpg the *E. coli* Fpg is a bilobal protein with a wide negatively charged DNA-binding groove.

Highly conserved residues Lys-57, His-71, Asn-169 and Arg-259 are involved in binding the phosphodiester backbone of DNA, which is sharply kinked at the lesion site. Conserved residues Met-74, Arg-110 and Phe-111 are inserted into DNA helix to fill the void in DNA after nucleotide eversion. A deep hydrophobic pocket in the active site is positioned to accommodate the damaged base.

BASE EXCISION REPAIR IN EUKARYOTES

Yeast

Monofunctional DNA glycosylase: spMYH: The *E. coli* MutY homolog was identified in *Schizosaccharomyces pombe* by a homology search in genome databases. The spMYH gene encodes for a 461 amino acids protein which is highly homologous to MutY and human MYH. spMYH possesses the HhH motif and the [4Fe-4S] DNA binding cluster domain found in Nth/MutY.

It has a DNA glycosylase and probably an AP lyase activity. SpMYH incises A opposite G (A/G), A/8-oxoG, 2-aminopurine/G and A/2-aminopurine containing DNA. Cells deleted for the spMYH gene (*spMyH*) have a mutator phenotype and are more sensitive to oxidative agents, such as hydrogen peroxide, than wild type cells.

These data show that spMYH is implicated in the avoidance of 8-oxoG induced mutations and it has an important role in the protection against oxidative stress. Interestingly, similar to human MYH (hMYH), spMYH also interacts directly with PCNA. This interaction is fundamental for the biological function of spMYH in the control of mutation avoidance since the mutation rate of *spMyH* expressing hMYH is reduced but it stays unchanged when *spMyH* cells express a mutant hMYH protein unable to interact with PCNA. *Bifunctional DNA glycosylases* S. cerevisiae *yOGG1, Ntg1 and Ntg2: Saccharomyces cerevisiae* 8-oxoguanine-DNA glycosylase (yOGG1) is the yeast counterpart of the *E. coli* Fpg protein. It was identified by functional suppression of the *E. coli fpg mutY* mutator phenotype, leading to the concept that eukaryotes use the DNA glycosylase pathway to repair oxidised purines in DNA. Comparison of amino acid sequences of the yOGG1 protein and *E. coli* Fpg protein does not reveal any obvious homology. The yOGG1 protein has associated DNA glycosylase/AP lyase activity and cleaves abasic sites via -elimination. In contrast to the bacterial enzyme, the yOGG1 protein repairs 8-oxoG only when paired with pyrimidines.

The yOGG1 also removes Fapy guanine-derived residues from DNA but less efficiently than 8-oxoG and -elimination was not observed. The major difference between Fpg and yOGG1 in terms of substrate specificity is that Fpg also removes adenine-derived Fapy lesions and oxidatively damaged pyrimidines. The yOGG1 is not an essential gene, as its disruption had no effect on the viability of a *ogg1* haploid mutant but *ogg1* cells display a mutator phenotype characterised by an augmentation of G/CT/A transversions. Like *fpg* cells, *ogg1* haploid mutants do not show particular sensitivity to H_2O_2. These data suggest that yOGG1 is the functional homologue of *E. coli* Fpg.

Based on the fact that the helix-hairpin-helix (HhH) DNA-binding motif was found in many DNA binding proteins, searches in databases for sequences containing this motif led to identification of the NTG1 gene on chromosome I of *S. cerevisiae*. The NTG1 gene encodes a 45 kDa protein, Ntg1p, which is homologous to the *E. coli* endonuclease III (Nth) but lacks the (4Fe-4S) cluster DNA binding domain found in the other members of this family.

Ntg1p is a DNA glycosylase/AP lyase which removes thymine glycols and unexpectedly releases formamidopyrimidine residues from DNA with a high efficiency comparable to that of the *E. coli* Fpg protein. Ntg1p also excises other lesions generated by oxidative stress such as 5,6-dihydrouracil, and 5-hydroxy-purines and 8-oxoG only when paired with guanine. Targeted disruption of the NTG1 gene results in viable cells. The Ntg1 protein localises both in the nucleus and in the mitochondria and is induced by cell exposure to DNA-damaging agents purified and characterised a second *S. cerevisiae* Ogg-like protein, called Ogg2, which preferentially acts on 8-oxoG/G base pairs. Later it turned out that Ogg2 is identical to Ntg1.

NTG2 was identified during analysis of *S. cerevisiae* genome as a second gene homolog to *E. coli* Nth in yeast. The NTG2 gene is localised on chromosome XV and codes for a 43.7 kDa protein, Ntg2p, which contains (4Fe-4S) cluster DNA binding domain also present in the *E. coli* Nth protein. The substrate specificity of Ntg2p is similar but not identical to Ntg1p; it cannot incise 8-oxoG mispaired with any of the four DNA bases. Recent results show that Ntg2p, but not Ntg1p, is implicated in the repair of certain degradation products of 8-oxoG such as 8-hydroxydeoxy-guanosine (8-OH-dG).

The targeted disruption of NTG2 results in viable cells but, similarly to *ntg1⁻* cells, greatly increases the rate of spontaneous and H_2O_2-induced mutations. Ntg2p is a nuclear enzyme constitutively expressed in cells. Interestingly, Ntg2p interacts with the DNA mismatch repair protein Mlh1p. Ntg1p and Ntg2p are both required for repair of spontaneous and induced oxidative DNA damage.

S. cerevisiae *AP endonucleases Apn1 and Apn2:* Homologs of *E. coli* Xth and Nfo have been identified in eukaryotes. The yeast APN1 gene encodes AP endonuclease I (Apn1) homologous to *E. coli* Nfo protein. This 41.4 kDa protein is the major AP endonuclease in yeast, accounting for >90% of total AP endonuclease activity. Apn1 has AP endonuclease, 3'-diesterase and 3'-phosphatase activities.

Like the *E. coli* homolog, Apn1 is a metalloenzyme excising 3'-phosphoglycoaldehyde, 3'-phosphoryl groups, and 3'-, unsaturated aldehydes. Yeast mutant lacking Apn1 (*apn1*) is viable but it is hypersensitive to both oxidative (H_2O_2 and t-butylhydroperoxide) and alkylating (methyl- and ethylmethane sulfonate) agents, it has 6- to 12-fold higher rate of spontaneous mutation than wild-type. This mutator phenotype is mainly characterized by a 60-fold increase in A/T to G/C transversion rate. The second yeast AP endonuclease (Apn2/ETH1) was identified by blast homology search with the sequences of *E. coli* Xth and human Ape1 in the *S. cerevisiae* genome.

Apn2 is a 520 amino acid protein, its expression (at the mRNA level) is induced by DNA damage. Apn2 has AP endonuclease, 3'-phosphodiesterase and 3'5' exonuclease activities, it can remove 3'-phosphate and 3' phosphoglycolate termini produced by H_2O_2. Interestingly, the 3' phosphodiesterase and the 3'5' exonuclease activities of Apn2 are 30-40-fold more active than its AP endonuclease activity. Apn2 could be an important factor in the repair of oxidative DNA damage. Yeast lacking apn2 (*apn2*) alone are viable and do not show a particular phenotype.

However *apn1 apn2* cells are remarkably sensitive to DNA damaging agents such as H_2O_2, MMS and phleomycin D1 and they present an increase of spontaneous mutation rate. Apn2 contains a carboxy-terminal domain which is absent in the other members of the family (Xth/Ape1). Deletion of this domain does not affect the enzymatic activity of Apn2 *in vitro* but the truncated protein cannot remove AP sites *in vivo*. These data strongly suggest that this domain is indispensible for protein-protein interaction and that Apn2 protein functions *in vivo* as a part of multiprotein complex.

Interestingly, it has been shown that the human homolog of Apn2 protein, Ape2, interacts with PCNA.

MAMMALS (HUMAN, RODENT)

Bifunctional mammalian DNA glycosylases OGG1, NTH1 and NEH1 hOGG1: Several groups reported the molecular cloning of the cDNA of the human and murine 8-oxoguanine DNA glycosylase (OGG1). The amino acid sequence of hOGG1 showed 33% identity and 54% similarity to *S. cerevisiae* OGG1 and conserved HhH/PVD motif which is present in endoIII/ MutY/AlkA superfamily of DNA glycosylases. The human OGG1 gene, located on chromosome 3, codes for seven alternatively spliced forms of mRNA. These transcripts were classified as type 1 and 2 depending on their last exon. The main type 1 and 2 transcripts encode, respectively, a 36 kDa nuclear and a 40 kDa mitochondrial polypeptide. These two proteins differ by their C-terminus. Both forms possess the same catalytic activity. Purified hOGG1 protein, similarly to *E. coli* Fpg protein, acts as a DNA glycosylase towards duplex DNA containing 8-oxoG/C base pair, Fapy and has associated AP lyase activity.

Crystal structure data at 2.1 Å resolution of the core domain of hOGG1 complexed with a 8-oxoG/C containing oligonucleotide was reported by Verdine's laboratory. This structure reveals that hOGG1 recognizes in the DNA helix both 8-oxoG (using amino acids F319, Q315, G42, C253) and the cytosine opposite 8-oxoG (using amino acids N149, R154, R204, Y203). The modified base is fully extruded from the helix and inserted into an extra-helical active-site pocket on the enzyme.

Interestingly, mutations of residue, which recognises the cytosine, does not alter the activity on 8-oxoG/C but significantly increases the activity toward 8-oxoG opposite other bases than cytosine. Physiologically, such mutants become both pro-mutagenic by repair of 8-oxoG/A and anti-mutagenic by repair of 8-oxoG/C. Study of search intermediates of hOGG1 protein in the presence of high molecular weight DNA by atomic force microscope shows that enzyme scans DNA at undamaged sites by inducing drastic kinks.

The base excision repair pathway for 8-oxoG involves the resynthesis of a single nucleotide at the lesion site as shown by *in vitro* repair assays with mammalian cell extracts and more recently by reconstitution *in vitro* of this pathway by using human purified proteins. These data show that the BER short-patch pathway is predominant for the repair of major oxidative DNA damage, 8-oxoG. Activity of hOGG1 is greatly stimulated by Ape1 which binds to AP sites generated by the DNA glycosylase and enhances the enzymatic turnover. This observation suggests that Ape1 might preclude the AP lyase activity of OGG1. The finding that the dRP lyase activity of DNA polymerase is required in repair of oxidative lesions by mammalian cell extracts supports this mechanism. The nuclear hOGG1 is associated with chromatin and the nuclear matrix during interphase and with condensed-chromatin during mitosis.

The fraction of hOGG1 bound to chromatin is phosphorylated on a serine residue, and the kinase responsible for this post-translational modification of hOGG1 might be protein kinase C. Homozygous *ogg1-/-* null mice were generated by targeted disruption. These mice are viable and, despite an increase of potentially mutagenic DNA lesions in their genome and mitochondria, they do not display any particular phenotype.

The *ogg1-/-* null mice have an elevated spontaneous mutation rate in non- or slowly proliferating tissues with high oxygen metabolism, such as liver, but they do not develop malignancies. These data are at variance with several reports which associate mutations and polymorphism of the OGG1 gene, in particular the Ser326Cys polymorphism, with increased risk of cancer.

Although the Ser326Cys polymorphism is slightly, or not associated with altered OGG1 activity, it might affect phosphorylation at this serine residue leading to adverse effects.

hNTH1: A human homolog of the *E. coli* Nth protein (endonuclease III) has been purified and cloned. This gene, called hNTH1 (human Nth homolog 1), encodes a 34.3 kDa protein that shares extensive homology with Nth including the conserved HhH-motif and the (4Fe-4S) cluster loop motif. hNTH1 protein acts as DNA glycosylase with an associated AP lyase activity, and has substrate specificity similar to the *E. coli* homolog towards damaged pyrimidine derivatives that result from ring saturation, ring fragmentation, ring contraction and unexpectedly ring-opened purine residues (Fapy). hNTH1 acts preferentially on 5-hydroxycytosines and AP sites when they are situated opposite to guanine and removes 8-oxoG opposite G. hNTH1 seems to be an exclusively nuclear protein and its expression is regulated during the cell cycle, showing a maximum in S-phase.

Homozygous *mNth1-/-* mutant mice do not have any detectable phenotype defect. Sensitivity of *mNth1-/-* mutant embryonic cells to H_2O_2 and menadione was not changed as compared to wild-type cells. Moreover, thymine glycol induced by ionising radiation is repaired, though more slowly than in wild type.

The absence of a significant phenotype could be explained by the discovery of two novel thymine-glycol DNA glycosylases in *mNth1* mutant cells called TGG1 and TGG2. These ~40 kDa proteins localise in mitochondria and nucleus, respectively. Thus, these data suggest the existence of several back-up DNA glycosylases in mammals to remove thymine glycol. The redundancy of such enzymatic activity highlights the importance of BER pathway for counteracting oxidative pyrimidine damage.

Although there is no definite evidence that BER enzymes act as multiprotein complexes, several reports suggest that addition of proteins can stimulate damage recognition and lesion processing. The NER enzyme XPG stimulates the hNTH1 activity *in vitro*. These data support the hypothesis that neurodegeneration in XP-patients could be due to defects in the repair of oxidative DNA damage. A yeast two-hybrid screen for other interacting partners of Nth1 led to the isolation of the DNA-binding protein B (DbpB)/Y box binding protein (YB-1). YB-1 greatly stimulates the hNTH1 activity. Altogether these data suggest that the BER pathways could be regulated by a number of non-BER proteins.

hNEH1: Human homologs of *E. coli* Nei/MutM were identified by database search of the genome and named hNEH1 and 2 (human Nei Homolog 1 and 2). hNEH1 was biochemically characterized, it encodes a 44 kDa protein which excises Fapy, oxidised purines and 8-oxoG when present in DNA. hNEH1 shows a tissue-specific expression with an S-phase specific increase at both RNA and protein level. Tissue-specific mRNA levels of OGG1 and NEH1 are distinct. These data suggest that the activities of OGG1 and NEH1 are not redundant at the tissue level and NEH1 might be involved in the replication-coupled repair of oxidative DNA damage.

Monofunctional DNA glycosylases ANPG, hTDG, MYH

ANPG: The ANPG protein is the human counterpart of the *E. coli* AlkA protein. This protein releases 3-meA and 7-meG from methylated DNA, but it is also able to excise hypoxantine and 1,N^6-ethenoadenine more efficiently than the AlkA protein. In contrast to the AlkA protein, the ANPG protein does not release 5-formyluracil and RT residues from DNA, however, there are activities in human cell-free extracts that can excise these lesions.

It should be noted that the human protein does not share significant amino acid sequence homology with the bacterial AlkA. Recently, it has been shown that ANPG efficiently excises 1,N^2-ethenoguanine (1,N^2-G) when present in DNA. Immunofluorescent staining for the ANPG protein in normal breast cells showed its nuclear localisation. Two alternatively spliced transcripts of human ANPG have been isolated from human cells.

The gene encoding for ANPG maps to chromosome 16. The full-length cDNA first isolated by Samson and colleagues (ANPG70, 293 AA, 1991) differs from the splice variant only in the sequence of the first 8 and 13 N-terminal amino acids respectively. Two truncated versions of human ANPG have also been described: ANPG40 lacks the first 63 amino acids and ANPG80 the first 73 amino acids from the N-terminus, when compared to APNG70. It has been concluded that the non-conserved, N-terminal part apparently contributes little to their damage recognition and N-glycosylase activity.

Surprisingly, it has been shown that the ANPG80, which lacks 73 amino acid residues at the N-terminus, is unable to excise 1,N^2-G from the duplex oligonucleotide and that the ANPG40 has a reduced activity as compared to the ANPG70. These observations indicate that the non-conserved, N-terminal part of ANPG is essential for 1,N^2-G-glycosylase activity but dispensable for the release of A, hypoxanthine and N-methylpurines. The ANPG activity toward hypoxanthine-containing substrates is slightly activated in the presence of the hHR23 protein.

Although the core part of the ANPG70 and APDG

proteins display 85% sequence identity, it seems that human enzyme repairs 1,N^2-G somewhat more efficiently than its rat counterpart. In fact, a similar difference between human and mouse ANPG in the recognition of 3-methylguanine and 7-methylguanine residues has been reported.

Interestingly, the human full-length and splice variants of ANPG have a direct repeat in N-terminal amino acid sequence, which is absent in the rat and murine homologues. These observations further support the role of the non-conserved N-terminal part of mammalian ANPG in substrate specificity.

To assess the role of this repair enzyme *in vivo,* mice deficient in 3-meAde-DNA glycosylase (*APNG/Aag* null mice) have been generated. Unexpectedly, these APNG knockout mice exhibit neither any particular sensitivity to alkylating agents nor any significant increase in the spontaneous mutation rate except for splenic T lymphocytes. Furthermore, Roth and Samson have shown that *Aag-/-* myeloid progenitor bone marrow cells are more resistant than wild-type cells to alkylating agents. This result suggests that the initiation of base excision repair could be more lethal to the cell than leaving the damaged bases unrepaired.

The crystal structures of ANPG80 (a truncated form of the ANPG protein) complexed to a DNA duplex containing pyridine and A have been established. ANPG80 is a single domain protein of mixed/ structures, which forms a distinct structural group that does not resemble any other BER protein. The enzyme bends the DNA by about 20° and intercalates into the minor groove of DNA causing the abasic pyrrolidine and A to slip into the enzyme active site. The structure of the ANPG80 nucleotide-binding pocket changes little upon flipping in the A base. The structure of the base-binding pocket of ANPG reveals the lack of specific interactions for methylated bases. This fact probably contributes to its remarkable multifunctionality, providing glycosylase activity within the same enzyme for deamination, methylation and exocyclic base damage.

hTDG: The human thymine DNA glycosylase (hTDG)

was first biochemically characterised for its ability to remove T mispaired with G. A more complete characterisation shows that hTDG exhibits a broad substrate specificity: it can excise T from DNA not only when mispaired with G but also from other T-containing mispairs except T/A.

Moreover, the hTDG protein more efficiently removes U from a G/U but not from a U/A mispair and 3,N^4-ethenocytosine opposite to all four natural bases. The hTDG gene is localised on the chromosome 12, it codes for several mRNAs which were cloned. These cDNA encode a single 46 kDa polypeptide (410 amino acids) which is homologous to the *E. coli* MUG protein. The hTDG protein has a high affinity for AP site and stays tightly bound to it after removal of T from a G/T mispair.

Ape1 activates hTDG by increasing the dissociation rate of hTDG from AP site probably through direct interaction. The stimulation of hTDG turnover by Ape1 is responsible for the efficient processing of 3,N^4-ethenocytosine by hTDG. A second factor facilitating hTDG enzymatic turnover was recently identified. Hardeland have shown that SUMOylation of hTDG by SUMO-1 and SUMO-2/3 drastically reduces its affinity for AP sites, increases its enzymatic turn-over towards G/U and reduces its processing activity on G/T substrate. SUMOylation also enhances the stimulatory effect of Ape1 on hTDG.

Using the yeast one-hybrid screen, hTDG was found to interact with retinoic acid receptor (RAR) and retinoic X receptor (RXR) which are ligand-dependent transcription factors. Interaction between RXR, RAR/RXR and hTDG enhances the binding of the former on their responsive elements. These data suggest that hTDG has a dual function: repair enzyme and transcriptional activator. hTDG is also a transcriptional repressor since it interacts with the thyroid transcription factor 1 (TTF-1) and then strongly inhibits the expression of TTF-1 responsive genes.

Recently, hTDG has been found to interact with the transcriptional co-activator CBP/p300. The hTDG/CBP/p300 complex is competent for both BER and histone acetylation.

hTDG enhances the CBP/p300 transcriptional activity and these proteins acetylate it. hTDG acetylation regulates the recruitment of Ape1 by hTDG. These data highlight the complexity of repair processes that are coupled to transcription, tightly regulated by post-translational modifications (SUMOylation and acetylation) and by protein-protein interactions.

SMUG1: The single-strand mismatch-specific uracil-DNA glycosylase (SMUG1) has been identified only in vertebrates by expression cloning. SMUG1 is a nuclear protein of 31 kDa which in fact efficiently removes uracil from double stranded DNA when paired with A and G, it also removes U from single-stranded substrates but less efficiently. Accumulation of 5-hydroxymethyluracil is associated with a cancer risk. The mammalian 5-hydroxymethyluracil DNA N-glycosylase activity is associated with SMUG1.

Like several DNA glycosylases, SMUG1 has a low turnover rate that could be due to its inhibition by AP sites. This could explain the Ape1-dependent stimulation of the hSMUG1 activity on both single and double-stranded DNA substrates. Targeted disruption of UNG gene in mice (*ung-/-*) does not result in relevant increase of spontaneous mutation rates, contrary to bacteria and yeast *ung*$^-$ mutants. It seems that hSMUG1 has an important role in the maintenance of genome stability since it represents the major uracil-DNA glycosylase activity in *ung-/-* cells.

hMYH: The human adenine-DNA glycosylase activity, called hMYH protein (human MutY homolog) was first purified from cellular extracts. Later using homology search, the genomic DNA and cDNA encoding hMYH were identified. Functional expression of hMYH in *E. coli* complements the mutator phenotype of *mutY* cells. The hMYH gene is localised on chromosome 1. It encodes a 59 kDa monofunctional DNA glycosylase which can excise A when opposite to 8-oxoG and to a lesser extent opposite to G. hMYH, similar to OGG1, is able to excise 2-OH-A, a mutagenic oxidative adduct. hMYH interacts with several proteins involved in the long-patch repair pathway such as PCNA, Ape1 and RPA.

In addition, hMYH interacts with the mismatch repair proteins hMSH2/hMSH6. Importantly, this interaction enhances the hMYH activity toward A/8-oxo-G. hMYH DNA glycosylase activity is equally enhanced by Ape1 on A/8-oxoG substrates. Unexpectedly this stimulation is independent of Ape1 AP endonuclease activity. Instead, Ape1 acts by stimulation of the formation of hMYH/DNA complexes. These data suggest that different repair pathways may cooperate through protein/protein interactions when counteracting oxidative DNA damage.

Ten different forms of mRNAs coding for hMYH were isolated. These mRNA encodes three proteins of 52, 53 and 57 kDa respectively. The 52 kDa and 53 kDa forms localise in the nucleus and the 57 kDa form in the mitochondria. Expression of hMYH is cell cycle-regulated with a maximum in S phase. Nuclear forms of hMYH co-localise with PCNA at DNA replication foci, suggesting a role of hMYH in replication-associated base excision repair.

Recently biochemical evidence has been provided that human cell extracts perform base excision repair of 8oxo-G/A mismatches on both strands. Similarly to what is reported in yeast, following adenine excision a cytosine is preferentially inserted opposite 8oxoG. This is followed by excision repair of 8oxoG in 8oxoG/C pair.

Interestingly, repair synthesis on either strand is completely inhibited by aphidicolin, suggesting that a replicative DNA polymerase is involved in the gap filling reaction. DNA polymerases/ are likely to be involved in this replication-associated BER. This was confirmed *in vivo* using murine MYH-deficient cells. *MYH-/-* cells repair A/8oxoG mispairs inefficiently but expression of wild-type MYH in these cells increases the repair. Interaction with PCNA is critical, since expression of a functional MYH lacking its PCNA-binding domain had no effect on the repair efficiency of A/8oxoG in *MYH-/-* cells. It has been speculated that hMYH might be considered as a cancer predisposing gene in humans since several mutations in this gene are linked with somatic mutations in *adenomatous polyposis coli* gene (APC).

AP endonucleases Ape1, Ape2: The major human AP endonuclease (HAP-1/Ape1/Apex), homologous to *E. coli* Xth protein (exonuclease III), was independently discovered as an AP-endonuclease and as redox-regulator of the DNA binding domain of Fos-Jun, Jun-Jun, AP-1 proteins and several other transcription factors including NF-kappa B, Myb and members of the ATF/CREB family. Ape1 also activates. Ape1 is a 35.5 kDa nuclear protein, which shows sequence homology to the *E. coli* Xth protein. Targeted homozygous disruption of Apex gene (*Apex-/-*) results in embryonic lethality in mice.

The heterozygous *Apex+/-* mice, so far, did not reveal significant differences in phenotypes associated with oxidative stress as compared to the wild-type. Due to the dual role of Ape1, the early embryonic lethality could be associated either with the BER defect or with the defective regulation of several transcription factors.

Ape1 can substitute for exonuclease III in *E. coli* and for Apn1 in *S. cerevisiae*). Beside its AP endonuclease activity, it exhibits other enzymatic activities: 3'5' exonuclease, phosphodiesterase, 3' phosphatase and RNase H. It should be noted that these additional activities are 3-4-fold weaker than its AP endonuclease activity. In addition, Chou and Cheng have shown that Ape1 has a 3'-mismatch exonuclease activity and so it might be considered as a proof-reading enzyme.

Ape1 plays a central role in BER. It is implicated in both short-patch and long-patch repair pathway. It could be also an important regulator of BER. Ape1 interacts with DNA polymerase *in vivo* which is a key element for BER, this interaction permits the loading of pol on DNA at the AP site and the enhancement of its dRPase activity. Ape1 interacts with other BER proteins such as the scaffold protein XRCC1, which stimulates its activity, and with PCNA and the Flap endonuclease 1 (FEN1).

The $p21^{WAF1/CIP1}$ is involved in the regulation of cell cycle progression, DNA replication and DNA repair. It can bind to PCNA inhibiting DNA replication and PCNA stimulation of long-patch BER pathway.

Ape1 seems to be implicated in the compensation of the

inhibitory effect of p21 on BER by stimulating and coordinating the long-patch BER. In addition, Ape1 is an activator of DNA glycosylases.

Altogether these data suggest that BER pathways are coordinated and regulated by protein-protein interactions and that Ape1 plays an important role in these interactions. Moreover, it has been shown that Ape1 and hnRNP-L are the components of the nCARE-B2 element binding complexes in the *ape1* gene promoter. This result suggests that expression of the *ape1* gene is down regulated by its own product. Ape1/ref-1 could also be involved in cell response to chemotherapy since it was found that nuclear accumulation of Ape1 is inversely proportional to that of p53 in head-and-neck cancer.

A second AP endonuclease, Ape2, homologous to the *S. cerevisiae* Apn2, was identified in mammals. The *ape2* gene is localised on the X chromosome and Ape2 transcripts are ubiquitously expressed. Ape2, a 57.3 kDa protein, exhibits only a weak ability to complement *E. coli* and *S. cerevisiae* AP endonuclease deficient cells. Ape2 localises predominantly in the nucleus but also in the mitochondria. In the nucleus, Ape2 co-localises with PCNA and interacts with it *in vitro*. These data suggest that Ape2 is implicated in both mitochondrial and nuclear BER and also that the nuclear role of Ape2 is PCNA-dependent.

OTHER ORGANISMS

Evidences have accumulated that oxidative DNA damage are repaired also through BER pathway in plants and insects. Homologues of the *Escherichia coli* Nth and Fpg proteins were characterized in *Arabidopsis thaliana*. Previously it was suggested that *Drosophila* has no BER pathway. However, an homologue of the human OGG1 protein in *Drosophila melanogaster* has been identified and characterized.

A new DNA glycosylase from *A. thaliana* ROS1 is a 1393 aminoacids protein, with an endonuclease III domain at the C-terminus. The recombinant ROS1 is able to incise oxidatively damaged DNA and also DNA containing 5 meC, suggesting a role in controling DNA methylation levels. This agrees with the pleiotropic developmental defects of the ros1 mutants suggesting regulatory properties of this protein. These results suggest that this protein could be a new class of DNA glycosylases, perhaps only present in plants..

ALTERNATIVE REPAIR PATHWAYS

So far, several alternatives to the BER repair pathways for oxidative DNA damage have been described.

Nucleotide Incision Repair

The BER pathway requiring sequential action of two enzymes for incision of DNA raises theoretical problems for the efficient repair of oxidative DNA damage because it generates genotoxic intermediates such as abasic (AP) sites and/or blocking 3'-termini groups that must be eliminated by additional steps before initiating DNA repair synthesis.

In addition, biological evidence hints at the existence of an alternative repair pathway. *E. coli* and mouse mutants deficient in the DNA glycosylases that remove oxidised bases are not sensitive to reactive oxygen species. A recent finding that Nfo and Nfo-like endonucleases nick DNA on the 5' side of various oxidatively damaged bases, generating 3'-hydroxyl and 5'-phosphate termini, provides an alternative pathway to the classic BER. It was proposed to name it the nucleotide incision repair (NIR) pathway.

The proposed NIR pathway eliminates the genotoxic intermediates, explains the genetic data and most likely identifies the physiological and original target as the modified base and not an artificial reduced abasic site, for the long patch repair pathway described in human cells. This alternative repair pathway is evolutionarily conserved from *E. coli* to human. The NIR pathway directly generates proper ends for DNA replication on one side and on the other side the site for elimination of the lesion by specific nucleases. This mechanistic characteristic creates an advantage as compared to BER.

NUCLEOTIDE EXCISION REPAIR

Oxidative DNA damage is a major source of genetic

mutation, and it is implicated in several human disorders including *Xeroderma pigmentosum* (XP). Human disorder XP is characterised by defect in nucleotide excision repair, the cells from XP patients fail to remove pyrimidine dimers caused by sunlight and, as a consequence, develop multiple cancers in areas exposed to light. Despite the fact that neural tissue is shielded from sunlight-induced DNA damage, a fraction of XP patients (~18%) display progressive neurologic degeneration secondary to a loss of neurones.

In fact, it was shown that in *E. coli* the nucleotide excision repair pathway could be involved in repair of oxidative DNA damage. Furthermore, in human cells two major oxidative DNA lesions, 8-oxoguanine and thymine glycol, are excised from DNA *in vitro* by the same enzyme system responsible for removing pyrimidine dimers and other bulky DNA adducts. These results support the idea that the NER pathway could be directly involved in repair of oxidative DNA damage.

Transcription-Coupled Repair (TCR)

Transcription-coupled repair (TCR) is a pathway of DNA excision repair that preferentially removes the DNA lesions present in transcribed sequences of expressed genes. TCR was originally observed for DNA damage repaired by NER pathway. However, recent studies demonstrated that Cockayne syndrome (CS) cells which are deficient for TCR cannot remove 8oxo-G in transcribed sequence, despite its proficient repair in non-transcribed sequence. Furthermore, it has been demonstrated that the breast and ovarian cancer susceptibility genes, BRCA1 and BRCA2, may participate in the repair of the 8oxo-G residues specifically located on the transcribed DNA strand in human cells.

Translesional DNA Synthesis

In *E. coli*, in addition to BER pathway, the translesion DNA synthesis system can contribute to the repair of chromosomal AP sites *in vivo*. In the last few years, new human DNA polymerases Pol-, Pol-, and Pol-, which are able to promote replication through DNA lesions were discovered. These polymerases catalyse translesion DNA synthesis (TLS) in a distributive manner and tend to show lower replication fidelity than other polymerases when unmodified DNA is used as a template. The yeast and human Pol- replicate DNA containing 8-oxo-G efficiently and accurately by inserting a cytosine across from the lesion. In support of the biochemical data, genetic studies show that synergistic increase in the rate of spontaneous mutations occurs in the absence of Pol- in the yeast *ogg1* mutant.

Both error-free and error-prone synthesis were observed in DNA synthesis across A residue catalysed by Pol- and Pol- . It has been found that Pol- preferentially misinsert G opposite to uracil thus providing a mechanism whereby mammalian cells can decrease the mutagenic potential of lesions formed via the deamination of cytosine. Altogether these observations suggest an additional role for TLS-specific DNA polymerases in the prevention of the mutagenic replication of oxidative DNA damage.

DNA Mismatch Repair (MMR) Pathway

DNA mismatch-repair system is involved in the repair of mispaired bases formed during replication, genetic recombination and as a result of DNA damage. Studies in *S. cerevisiae* indicate the involvement of the mismatch repair pathway in prevention of genotoxic effect of oxidative DNA damage. It was shown that mutations in MSH2 and MSH6 caused a synergistic increase in mutation rate in combination with mutations in OGG1, resulting in a 140- to 218-fold increase in the G/C-to-T/A transversion rate. Consistent with this, the MSH2-MSH6 complex binds to 8oxoG:A mispairs and 8oxo-G/C base pairs with high affinity and specificity.

Another important source of 8oxo-G in DNA is the dNTP pool. Recently it has been shown that incorporated 8oxo-dGMP contributes significantly to the mutator phenotype of MMR-deficient cells. Increased expression of MTH1 in MSH2-/- cells produces a significant decrease of DNA 8oxo-G levels and is associated with a drastic reduction of the mutation rate. These findings indicate that MMR excises incorporated 8oxoGMP from newly synthesised DNA strand, thus providing cells with an additional protective strategy from oxidative stress.

Homologous Recombination Repair Pathway

It is well established that homologous recombination (HR) system is pivotal for the repair of an important form of oxidative DNA damage induced by ionising radiation - the double-strand break (DSB). Moreover, *in vivo* replication forks often encounter template DNA damage that can lead to replication fork collapse generating double-strand break. It is hypothesised that DNA recombination functions as a system for reactivation of the collapsed replication fork. The importance of this function to the formation of lethal DNA damage is underscored by the uncontrollable fragmentation of chromosomes in Rad51-deficient vertebrate cell lines.

Another important characteristic of the HR repair pathway is that it mainly provides a non-mutagenic replicative bypass of the blocking lesion using sister chromatid or homologous chromosome as an undamaged DNA template. Recently, it has been found that defects in the homologous recombination pathway are associated with cancer-prone clinical syndromes, in particular *ataxia telangiectasia,* hereditary breast cancer, Bloom's syndrome and Werner's syndrome. These observations support the idea that the HR system could be an alternative to excision/incision repair pathways in counteraction of genotoxic effects of oxidative DNA damage.

Nucleotide Pool-sanitising Enzymes

Besides the modification of DNA by ROS, these species also oxidise the nucleotide pool. The modified nucleotides could be introduced in DNA and also in RNA during their synthesis leading to mutations. The *E. coli* mutT mutant shows a 100- to 10 000-fold increase in A/TG/C transversion rates as compared to wild-type. This effect is due to the accumulation of mutagenic 8oxo-GTP in the cellular dNTP pool which is incorporated opposite A and C during DNA synthesis, thus leading to A/TG/C transversions. The *E. coli* MuT protein is a 8oxo-GTPase that cleanses the cellular nucleotide pool from

8oxoGTP. In addition, MutT hydrolyses ribonucleotide 8oxo-GTP, suggesting the importance of this enzyme in maintaining transcription fidelity.

Mammalian cells contain MutT homolog, the MTH1 protein (MutT Homolog 1). The corresponding cDNAs were cloned in human and murine cells, and protein was characterised. MTH1 displays only 23% identity to the *E. coli* counterpart but the highly conserved region essential for the 8oxo-GTPase activity contains identical residues. Therefore, expression of MTH1 human and rodent origins in *E. coli mutT* reverts to the mutator phenotype.

These results indicate that mammalian proteins may have a similar antimutagenic activity *in vivo* as *E. coli* MutT protein. Like MutT, MTH1 hydrolyses 8oxo-GTP and oxidised ribonucleotides 8oxo-GTP. Interestingly, MTH1 possess an additional 2-hydroxyadenosine (2-OH-A) triphosphate pyrophosphorylase activity. Since hMYH (MutY homolog) is able to remove 2-OH-A from DNA, these data suggest that mammalian cells possess a pathway to sanitise oxidised forms of both guanine and adenine. Seven types of mRNA transcripts, coded by the MTH1 gene, have been identified in human Jurkat cells, which lead to the synthesis of, at least, three polypeptides of 18, 21 and 22 kDa. MTH1 polypeptides localise mainly in the cytoplasm and in mitochondria.

The MTH1 knockout cell lines and mice were constructed by gene targeting. MTH1-/- cells exhibit a twofold increase of mutation rate as compared to MTH1+/+ cells which is in contrast to the increase of mutation rate in *E. coli mutM* cells. Therefore, it seems likely that mammalian cells have other enzyme(s) and/or pathway(s) capable of hydrolysing oxidised nucleotides. MTH1-/- mice exhibit slightly elevated tumour incidence than wild-type mice but their survival at 1.5 years is not affected.

Several studies on human cancers show that mutations in hMTH1 gene, so far are not involved in these pathologies. Unexpectedly, MTH1 and translesional DNA polymerase kappa (Pol-) expressions were found to significantly increase and decrease respectively in two mammary carcinoma cell lines characterised by frequent A/TG/C transversions. So far, no human diseases have been linked to defects in proteins involved in the BER pathway. DNA repair genes functionally expressed in mammalian cells and now in transgenic mice having a null mutation in the gene coding for BER proteins are very important tools to ascertain the biological role of these proteins *in vivo*.

It has been very astonishing to notice that, beside a few examples of targeted deletion of genes encoding the AP-endonuclease 1 and DNA polymerase in mice leading to embryonic lethality, the genotype of the other BER protein knockout mice (such as a number of DNA glycosylases) does not show any striking particularity in terms of predisposition to cancer and premature aging.

Furthermore, examination of 6200 *S. cerevisiae* genes transcript levels after exposure to various genotoxic agents reveals that the DNA repair genes are only modestly induced, which is in contrast to adaptive and SOS response in *E. coli*. These observations suggest the possibility of back-up repair pathway(s) for oxidative DNA damage that have to be characterized. One could expect important breakthroughs from crosses between different strains to produce double knockouts to identify the possible back-up systems, the processes involved in the regulation and the interactions of the different pathways.

Detailed understanding of the mechanisms leading to the co-ordination of various proteins involved in the molecular reaction of BER is of paramount importance for gaining insights into the efficiency and fidelity of this key pathway for genome stability, prevention of cancer, resistance to chemotherapeutic agents, degenerative diseases and more recently in some aspects of teratogenicity.

TRANSCRIPTION FACTORS

Transcription factors play essential role in the gene regulation in higher organisms, binding to multiple target sequences and regulating multiple genes in a complex manner. In order to understand the molecular mechanism of target recognition, and to predict target genes for transcription factors at the genome level, it is important to analyse the relationship between the structure and function (specificity) of transcription factors.

We have used a knowledge-based approach, utilizing rapidly increasing structural data of protein-DNA complexes, to derive empirical potential functions for the specific interactions between bases and amino acids as well as for DNA conformation, from the statistical analyses on the structural data. Then these statistical potentials are used to quantify the specificity of protein-DNA recognition.

The quantification of specificity has enabled us to establish the structure-function analysis of transcription factors, such as the effects of binding cooperativity on target recognition. The method is also applied to real genome sequences, predicting potential target sites. We are also using computer simulations of protein-DNA interactions in order to complement the empirical method.

Combining the two approaches together, we can better understand the mechanism of protein-DNA recognition and improve the target prediction of transcription factors. Regulation of gene expression in higher organisms is achieved by a specific recognition of target DNA sequences by DNA-binding proteins. Due to the progress of X-ray crystallography and NMR spectroscopy techniques, structural data on the protein-DNA complexes have been rapidly increasing. However, the mechanism of DNA sequence recognition by proteins has been poorly understood, and thus the accurate prediction of their targets at the genome level is not yet possible. This situation implies that the structural information has not been fully utilized.

Understanding the molecular mechanism and its application to genomewide prediction are essential for the analysis of gene regulation network. Here, we describe two kinds of approaches for studying protein-DNA recognition. One is a knowledge-based approach, by which we extract functional information from structural data of protein- DNA complexes.

We made a statistical analysis of structural database of protein-DNA complex, and derived empirical potential functions for the specific interactions between bases and amino acids. Then, we used a sequence-structure threading to examine the relationship between structure and specificity in protein- DNA recognition, and to predict target sites of transcription factors in real genome sequences. We also evaluated the fitness of DNA sequence against DNA structure to examine the role of indirect readout mechanism.

We show how the structural features are related to the specificity, and discuss relative roles of direct and indirect readout mechanisms in the recognition. We also discuss the strategy to predict the target sequences of transcription factors at the genome level, and its application to yeast genome. In another approach, we analyse protein-DNA recognition by computer simulations. We describe several different methods of computer simulations.

RELATIONSHIP IN PROTEIN-DNA RECOGNITION

It is interesting to compare two structures, cognate and non-cognate complex structures in order to understand what is important for specific binding and what is different between them. There are several such examples in PDB (Protein Data Bank). We examined NF-kB, glucocorticoid receptor DNA binding domain, EcoRV endonuclease and BamHI endonuclease, for which both the cognate and non-cognate complex structures are available in PDB. The statistical method could distinguish the two structures as differences in the Zscores as well as statistical poteitials.

Thus, the subtle differences in specificity of these structures could be detected by the analysis of energies. Proteins often bind to DNA as homodimers, which leads to subtle structural differences between the two subunits. Thus, we examined in detail the structural effects of asymmetric binding on specificity. Marked differences in the specificity of DNA binding were observed for the two subunits of l repressor, the glucocorticoid receptor, and for transcription factors containing a Zn_2CyS_6 binuclear cluster domain, which are known to bind asymmetrically to DNA. We also applied this method to examine the relationship between structure and specificity in cooperative protein-DNA binding. The effect of cooperative binding was examined by comparing the monomer and heterodimer complexes of MATa1/a2, MCM1/ MATa2 and NFAT/AP- 1 transcription factors. We found that the heterodimer binding enhances the specificity in a non-additive manner.

This result indicates that the conformational changes introduced by the heterodimer binding play an important role in enhancing the specificity. The structures of cognate and noncognate complexes of EcoRV show marked differences in the conformations of both the enzyme and DNA. The example of EcoRV enabled us to show the cooperative effect of sequence and structure on specificity - i.e., conformational differences between the cognate and noncognate complexes were sensitive to the cognate but not the noncognate DNA sequence, and the cognate structure had a greater ability to discriminate DNA sequences than the noncognate structure.

The above method is based on the direct readout mechanism, in which protein recognizes DNA sequence through the direct contact between amino acids and base pairs. On the other hand, substitutions of those base pairs not in contact with amino acids often affect binding affinity, indicating that protein may recognize DNA sequence through particular structure or property of DNA. This indirect readout mechanism may contribute to the specificity of protein-DNA recognition significantly. We have derived statistical potential functions for conformational energy of DNA to quantify the specificity of indirect readout mechanism of protein- DNA recognition.

Once the potentials are derived, the conformational energy of DNA sequence can be estimated for given structure, and the threading procedure can be used to evaluate the fitness of sequence to structure of DNA. We can calculate the Z-score for the indirect recognition in the same way as for the direct recognition. By comparing both the Z-scores we can assess the relative contributions of direct and indirect readout mechanisms. We have analyzed various protein- DNA complexes systematically, and found that both the direct and indirect mechanisms make significant contribution to the specificity. The relative contributions depend on the types of DNAbinding proteins. Because both the potentials are independent quantities, they can be summed up to calculate the total energy and used to find target sites, although a weighting factor needs to be determined as the two potentials were derived from different statistics.

The Zscore calculated for the total energy with varying weighting coefficient. One can see enhanced Z-score compared with individual Z-scores for direct or indirect readout (c=1 or 0, respectively). This result indicates that the energies of the direct and indirect readouts contain independent information that in combination enhances the specificity of the recognition. We use the weight coefficient that gives the maximum total Z-score. We use the combined energy for threading against DNA sequences to make target predictions for transcription factors.

Prediction in Yeast Genome

The threading procedure was used to find target sites of transcription factors in real genome sequences. As an example of such applications, we could identify the experimentally-verified binding sites of the transcription factor MATa1/a2 in the promoter of *HO* gene successfully. We have also attempted to identify target sites and genes of MCM1/MATa2 in the whole yeast genome.

The target genes of this transcription factor have been identified in yeast genome experimentally. The predicted target genes were ranked by the Z-score and compared with experimental data. The target genes identified positively by experiment were ranked high in the list, and the experimentally negative genes were ranked low.

Separation between the positive and negative genes was not perfect but they were segregated by a certain threshold Z-score value. The total Z-score gave better separation than that of direct contribution alone. We are applying the method for the targets prediction of other transcription factors.

PROTEIN-DNA INTERACTIONS

The statistical method has shown that the distribution of Ca position of amino acids around base pairs provides important information about the specificity in the DNA sequence recognition by proteins. However, the accuracy of this prediction method is limited by the number of available structural data. Thus, it is desirable to complement the method by some other means. Computer simulation is one such method. In real DNA-protein interactions, however, there are many factors contributing to the recognition process. Thus, it is necessary to examine different levels of the system by different methods.

Free-Energy Map for Base-Amino Acid Interactions

In the case of base-amino acid interactions, we need to reproduce the distribution of Ca position of amino acids around base pairs observed in the experimental DNAprotein complex structures. Thus, it would be reasonable to consider at first the interactions between a base pair and an amino acid. In reality, the Ca position is fixed by main chain of protein and the possible range of Cb direction may be restricted. However, such biases, which are context dependent, are difficult to evaluate *a priori*. Therefore, at first we will consider intrinsic interactions between a base pairs and an amino acid.

We generated many Ca positions and side chain conformations by systematic sampling or Monte Carlo sampling methods, and calculated free energy map of Ca around a given base pair. By calculating the free energies for different Ca positions and subtracting a reference free energy at a large separation, we can obtain a contour map of interaction free energy, which shows preferable positions of Ca of amino acid around a base pair. This can be directly compared with the distribution of Ca position of amino acids around base pairs derived from DNA-protein complex database.

According to the freeenergy contour maps for the interactions of Asn with AT and with G-C, the preferable position of Ca is localized in a narrow region around A in the case of A-T. In this region, Asn and A form specific double hydrogen bonds, C=O...HN6 and NH...N7, which are found frequently in the Asn-A pair in the experimental structures of DNA-protein complexes. Also, the distribution of Ca is in agreement with the statistical potential obtained by the database analysis.

On the other hand, Asn tends to be more broadly distributed around G-C. The lowest DG values are located in the middle of G-C, and the following lower DG's extend towards C5 atom of C, where C does not have a methyl group. This comparison indicates that the interaction of Asn is more specific toward A-T than G-C. This example illustrates how we can quantify the specificity in the base-amino acid interactions and complement the statistical method.

CHANGES DUE TO BASE MUTATIONS

The specificity of protein-DNA recognition is usually assessed by the effect of mutations. If mutations cause large effect on the binding affinity, it would indicate that specific interactions are involved at the mutation site, and the extent of specificity dictates the effect of mutation. Thus, it is important to examine the effect of mutation and component of interaction energy causing the effect, as well as to develop methods to predict the mutation effect. We have performed a computer analysis in which a phage DNA-binding protein, l repressor, was used to examine the changes in binding free-energy (DDG) and its energy components caused by single base mutations. We then determined which of the calculated energy components best correlated with the experimental data. The experimental DDG values were well reproduced by the calculations.

Component-analysis revealed that the electrostatic and hydrogen bond energies were most strongly correlated with the experimental data. Among the 51 single basesubstitution mutants examined, positive DDG values, corresponding to weakened binding, were caused by the loss of favorable electrostatic interactions and hydrogen bonds, the introduction of steric collisions and electrostatic repulsion, the loss of favorable interactions with a thymine methyl group, and the increase of unfavorable hydration energy from isolated DNA. These results suggest that electrostatic interactions and hydrogen bond make major contributions to the specificity of l repressor-DNA recognition. However, van der Waals and hydrophobic interactions also play important role in particular DNA sequence locations.

PROTEIN-DNA COMPLEXES

For a large system like l repressor DNA complex, it is not an easy task to use all-atom simulations to calculate interaction free energy changes caused by mutations accurately. However, calculation of free energy changes due to small structural perturbation in the complex is feasible. We calculated a binding free energy change between DNA and l repressor due to a base substitution from tymine (T) to uracil (U) by the free energy perturbation method based on molecular dynamics simulations for the complex in water with all degrees of freedom and long-range Coulomb interactions.

The binding free energy change calculated was in good agreement with an experimental value. By using component analysis, we could clarify the reason why the small difference between T and U (CH3 in T is replaced by H in U) caused the significant binding free energy change. The free energy change is due to the gain of hydration free energy in the dissociated state and the loss of favorable van der Waals interactions in the associated state.

Docking Simulations of Protein-DNA Complexes

During the recognition process, proteins associate and dissociate with DNA, and may slide along DNA before finding their target sites. Thus, protein-DNA recognition is a dynamical process. In order to reveal reaction pathways of proteins along DNA, we are conducting docking simulations of protein-DNA complexes by using Monte Carlo method. At first, using rigid-body approximations, we examined free-energy profile along the DNA surface for homeodomains and restriction enzymes. Preliminary results indicate that the homeodomain may track along the major groove of DNA while it slides on DNA. We will incorporate the flexibilities of protein and DNA molecules such as hinge and bending motions, and examine the role of these effects on the protein-DNA recognition process.

We have described two kinds of methods for studying the relationship between structure and function (specificity) in protein-DNA recognition. One is the knowledge-based approach, by which we extract biological information from structural data on protein- DNA complexes. The increase in the structural data, which will be accelerated by structural genomics project, will make this structure-based method promising for revealing the structure-function relationship in protein- DNA recognition and for predicting targets of transcription factors. This method can also be applied to proteins of unknown structure having substantial sequence similarity to known proteins: the structure can be modeled based on the similarity and its binding sites can be predicted.

This approach is, however, limited by the number of available data. Thus, we try to complement it with a deductive approach, by which we try to reproduce the recognition process by computer simulation. Although this approach requires a large amount of computational time, the advancement of technology makes the computer simulation of large systems feasible. The information obtained by this approach would help the knowledge-based approach in improving the accuracy of statistical potentials and target prediction. Thus, a combination of the two methods will become a powerful tool for the study of protein-DNA recognition and its application to target prediction at the genome level.

Chapter 9

Types of Chromatin

Heterochromatin stains more strongly and is a more condensed chromatin. Euchromatin stains weakly and is more open (less condensed). Euchromatin remains dispersed (uncondensed) during Interphase, when RNA transcription occurs. Some regions of heterochromatin appear to be structural (as in the heterochromatin near the centromere region). Barr bodies, irreversibly inactivated X-chromosomes, are also condensed heterochromatin. Other heterochromatin regions vary from cell to cell.

As the cell differentiates, the proportion of heterochromatin to euchromatin increases, reflecting increased specialization of the cell as it matures. Loops (or puffs) in insect chromosomes are areas of active RNA synthesis, suggesting again the functional genes are located in open areas of the chromatin (or, the euchromatin).

Eukaryotes also have specific binding proteins working in a similar fashion to prokaryotic mechanisms, however the eukaryotes, as one would expect, have a much more complicated process.

THE EUKARYOTIC GENOME

We use the term genome to refer to all of the alleles possessed by an organism (or by a population, species, or larger taxonomic group). While the amount of DNA for a diploid cell is constant within a species, the differences can be great between species. Humans have 3.5×10^9 base pairs, *Drosophila* has 1.5×10^8, toads have 3.32×10^9, and salamanders have 8×10^{10} base pairs per haploid genome. Much of the

DNA in each cell either has no function or has a function not yet known. Eukaryotes have only 10% of their DNA coding for proteins. Humans may have a little as 1% coding for proteins. Viruses and prokaryotes use a great deal more of their DNA.

Almost half the DNA in eukaryotic cells is repeated nucleotide sequences. Protein-coding sequences are interrupted by non-coding regions. Non-coding interruptions are known as intervening sequences or introns. Coding sequences that are expressed are exons.

Most, but not all structural eukaryote genes contain introns. Although transcribed, these introns are excised (cut out) before translation (a seemingly energy inefficient process). The number of introns varies with the particular gene, even occurring in tRNAs, rRNAs and viral genes! Generally the more complex and recently evolved the organism, the more numerous and larger the introns.

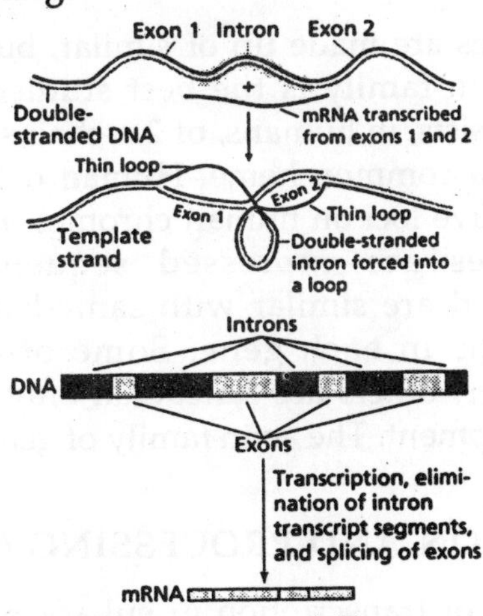

Fig. Introns and Exons

Which came first: continuous genes lacking introns or interrupted genes containing introns? Introns have been hypothesized to promote genetic recombination (via crossing-over), thus speeding up the evolution of new proteins. Exons are also thought to code for different functional regions of proteins.

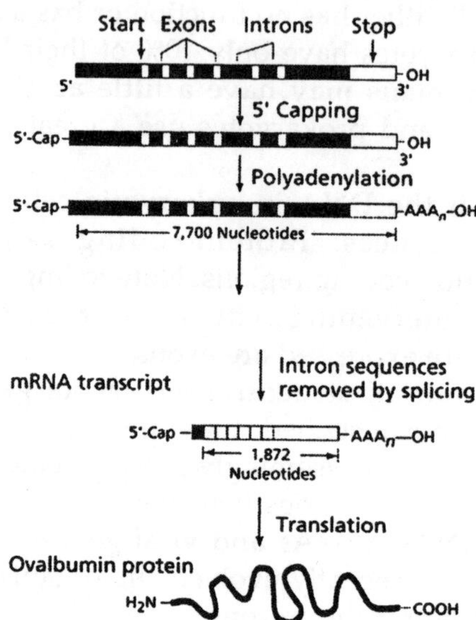

Gene families are made up of similar, but not identical, genes. The globin family is the best studied gene family. Hemoglobin consists, in humans, of 2 a-chains and 2 b-chains clustered about a common heme. Human beta-globin genes are scattered at five loci on human chromosome 11.

These genes are expressed sequentially during development, and are similar with same-length introns in similar positions in each gene. Some of the genes are inactivated copies, others are functional only during certain phases of development. The actin family of genes also exhibits a similar pattern.

TRANSCRIPTION AND PROCESSING OF mRNA

The process of transcription in eukaryotes is similar to that in prokaryotes, although there are some differences. Eukaryote genes are not grouped in operons as are prokaryote genes. Each eukaryote gene is transcribed separately, with separate transcriptional controls on each gene. Whereas prokaryotes have one type of RNA polymerase for all types of RNA, eukaryotes have a separate RNA polymerase for each type of RNA.

One enzyme for mRNA-coding genes such as structural proteins. One enzyme for large rRNAs. A third enzyme for smaller rRNAs and tRNAs. Prokaryote translation begins even before transcription has finished, while eukaryotes have the two processes separated in time and location (remember the nuclear envelope). After eukaryotes transcribe an RNA, the RNA transcript is extensively modified before export to the cytoplasm. A cap of 7-methylguanine (a series of an unusual base) is added to the 5' end of the mRNA; this cap is essential for binding the mRNA to the ribosome.

A string of adenines (as many as 200 nucleotides known as poly-A) is added to the 3' end of the mRNA after transcription. The function of a poly-A tail is not known, but it can be used to capture mRNAs for study. Introns are cut out of the message and the exons are spliced together before the mRNA leaves the nucleus. There are several examples of identical messages being processed by different methods, often turning introns into exons and vice-versa. Protein molecules are attached to mRNAs that are exported, forming ribonucleoprotein particles (mRNPs) which may help in transport through the nuclear pores and also in attaching to ribosomes.

ANTIBODY-CODING GENES

Antibodies are globular complex proteins made by multicellular individuals in response to a specific antigen (a foreign substance that has labels saying "I am a stranger"). The cells making antibodies are lymphocytes, better known as white blood cells. Antibodies immobilize and destroy their specific antigens. A mouse can make 10,000,000 antibodies, more than the total genes of the mouse would suggest. Antibody proteins are composed of two long and two short chains. Each species has a constant region characteristic for the species and type of antibody. The other region is the variable in which antigen-specific amino-acids are located.

Susumu Tonegawa tested (and proved) an old hypothesis that the constant and variable regions of antibodies were coded for by different genes. Tonegawa's work further demonstrated that gene fragments in embryos are rearranged to form the variable (functional) genes.

VIRUSES AND EUKARYOTES

The viruses of eukaryotes are similar to prokaryote-infecting viruses. Proviruses are viral DNA integrated into the host cell. Some of the DNA viruses can either initiate an infection (lytic in prokaryotes) cycle or can form proviruses. Simian Virus 40 (SV40) causes cancer in hamsters but not in its normal hosts. SV40 can introduce new functional genes into the host DNA, as can a number of other viruses.

Retroviruses can also insert their nucleic acid into host DNA via the reverse transcriptase mode mentioned previously. Reverse transcriptase is carried with the RNA into the host cell. In the process of making the cDNA strand for the viral RNA, the enzyme also makes long terminal repeats (LTRs) sequences at the terminal ends of the cDNA. These LTRs may also make insertion of the viral DNA into the host DNA easier. The inserted viral DNA makes RNA transcripts which are packaged with viral protein coats and reverse transcriptase.

The viral DNA may, depending on its insertion point, cause mutations of the host DNA. Most viral DNA insertions do not damage the host, but rather become part of the host genome and can be passed on if they have managed to infect a germ-line cell. From 0.5 to 1.0% of mouse DNA may be of viral origin.

EUKARYOTIC TRANSPOSONS

Eukaryotic transposons resemble prokaryotic transposons in many features. Many eukaryotic transposons are first copied to RNA and then back to DNA before being inserted in a new location.

GENES, VIRUSES AND CANCER

Cancer is a disease in which cells escape the restraints on normal cell growth. Cancer is an inheritable disease (at least from cell to daughter cells). Once a cell has become cancerous, all of its descendant cells are cancerous. Gross chromosomal abnormalities are often visible in cancerous cells. Most carcinogens (cancer-generating factors) are also mutagens (mutation-generating factors). Oncogenes are genes resembling normal genes but in which something has gone wrong, resulting in a cancer. Fifty oncogenes have thus far been discovered.

Viruses seem able to cause cancer in three ways. Presence of the viral DNA may disrupt normal host DNA functions. Viral proteins needed for virus replication may also affect normal host gene regulation. Since most cancer-causing viruses are retroviruses, the virus may serve as a vector for oncogene insertion. Transfers of genes between eukaryotic cells will allow doctors, who have historically been limited to phenotypic cures, to attack disease at the genotypic level. SV40 virus has been used to inject the rabbit beta-globin gene into monkeys. Viruses can thus serve as a possible vector to place healthy (non-mutated) alleles into eggs.

GENE REGULATION IN EUKARYOTES

Like prokaryotes, eukaryotic organisms do not want to express all of their genes all of the time. Given the complexity of multicellular eukaryotes, gene regulation in these organisms needs to be very complex. This module provides a brief overview of the various levels of regulation of gene expression in eukaryotes, and takes a look at the basics of transcriptional regulation; a detailed look at eukaryotic transcriptional regulation is beyond the scope of this course.

GENE REGULATION IN EUKARYOTES

Eukaryotes need to regulate their genes for different reasons than prokaryotes. In prokaryotes, gene regulation allowed them to respond to their environment efficiently and economically. While eukaryotes can respond to their environment, the main reason higher eukaryotes need to regulate their genes is cell specialization. Whereas prokaryotes are (relatively speaking) simple, unicellular organisms, multicellular eukaryotes consist of hundreds of different cell types, each differentiated to serve a different specialized function. Each cell type differentiates by activating a different subset of genes. Because of the multitude of cell types, the regulation of gene expression required to bring about such differentiation is necessarily complex. One way this complexity is demonstrated is in multiple levels of regulation of gene expression.

LEVELS OF REGULATION

Before we discuss the specifics of regulation, it is necessary to understand that "gene expression" covers the entire process from transcription through protein synthesis. The final measure of whether or not a gene is "expressed" is if the protein is produced, because it is protein that will ultimately carry out the function specified by the gene.

We've seen numerous examples of how eukaryotic cells are more complex than prokaryotic cells. One obvious example of this is the presence of a nucleus in eukaryotic cells, which separates transcription from translation in a way not seen in prokaryotes.

Furthermore, eukaryotic transcripts must be processed before they can be translated. Here is a diagram outlining the steps involved in the production of a protein in eukaryotic cells:

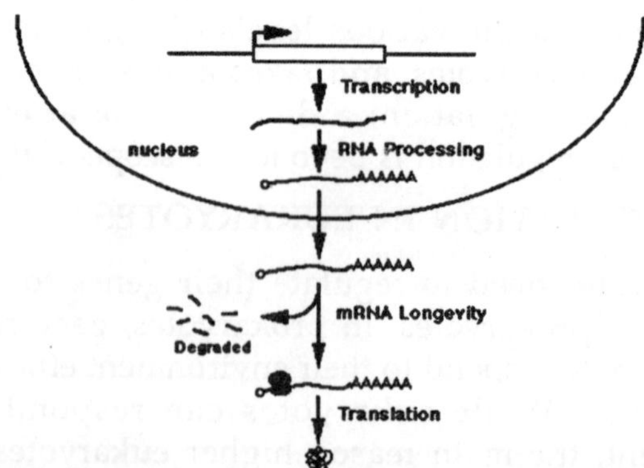

Regulation can occur at any point in this pathway; specifically, it occurs at the levels of transcription, RNA processing, mRNA lifetime (longevity), and translation. Each of these types of regulation will be considered in turn.

REGULATION OF RNA PROCESSING

After transcription, the RNA must be processed before it can be translated. As described elsewhere, RNA processing involves addition of a 5' cap, addition of a 3' poly (A) tail, and removal of introns. This processing represents another level of regulation of gene expression, particularly in regard to splicing out of introns. Regulation can be of two types:

- Whether an RNA gets processed;
- Which exons are retained in the mRNA.

The first type of regulation can determine whether or not an mRNA gets translated. If an RNA is not processed, it will not be transported out of the nucleus, and will not be translated.

The second type of regulation can affect the function of the protein produced. Some genes have exons that can be exchanged in a process known as exon shuffling. For example, a gene with four exons might be spliced differently in two different cell types. In cell 1, exons 1, 2, and 4 would be used in the mRNA:

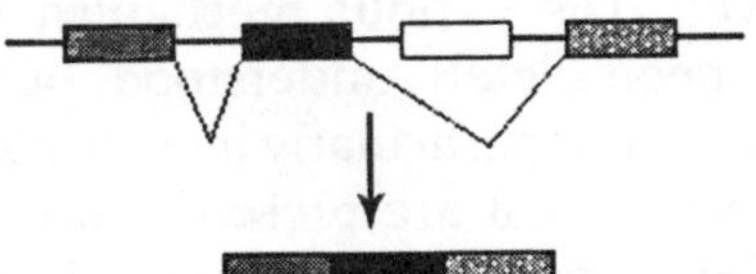

In cell 2 on the other hand, exons 1, 3, and 4 would be used:

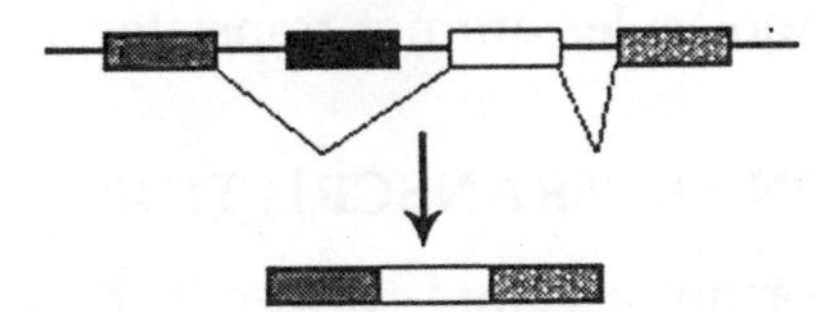

In each of these cases, the polypeptide produced could have a different function. In mammals, for example, the calcitonin gene produces a hormone in one cell type, and a neurotransmitter in another cell type, due to exon shuffling. In *Drosophila,* alternate splicing of the *sex-lethal* RNA can produce an mRNA encoding a functional polypeptide, or one with a premature stop codon that encodes a short, nonfunctional polypeptide.

REGULATION OF RNA LONGEVITY

Imagine two mRNA molecules: one lasts for five minutes in the cytoplasm before being degraded, while the other one manages to linger for an hour before being degraded. If both are translated continually while they exist, it is obvious that more of the second polypeptide will be produced than the first. This is the principle behind regulation of RNA longevity. mRNAs from different genes have their approximate lifespan encoded in them; this serves to help regulate how much of each polypeptide is produced. The information for lifespan is found in the 3' UTR. The sequence AUUUA, when found in the 3' UTR, is a signal for early degradation (and therefore short lifetime). The more times the sequence is present, the shorter the lifespan of the mRNA. Because it is encoded in the nucleotide sequence, this is a set property of each different mRNA; the longevity of an mRNA can't be varied.

REGULATION OF TRANSLATION

Whether or not an mRNA molecule is translated can be regulated as well. The various mechanisms of translational regulation are incompletely understood, but there are many documented examples (particularly in embryonic development) of mRNA molecules that are present routinely, but are only translated under certain circumstances. For example, many animals sequester large amounts of mRNA in their eggs, and those mRNA molecules are not translated unless the egg is fertilized.

REGULATION OF TRANSCRIPTION

Whether or not a gene is transcribed is the major way that gene expression is regulated in eukaryotes, as it was in prokaryotes. There are some major differences between transcriptional regulation in prokaryotes and eukaryotes. For one thing, because of the complexity of eukaryotic patterns of gene expression, each eukaryotic gene needs its own promoter. In other words, eukaryotic genes are not organized into operons. Another difference is that prokaryotic genes are regulated primarily by repressors. Although repressors occasionally play a role in eukaryotes, eukaryotic genes are primarily regulated by transcriptional activators. These activators are transcription factors.

ELEMENTS OF EUKARYOTIC GENES

As discussed in the module on transcription, eukaryotic genes have promoters that are recognized by basal transcription factors (such as TFIID). In addition to the promoter, eukaryotic genes have one or more enhancers. These are DNA sequences associated with the gene being regulated, and whereas the promoter is responsible for initiating low levels of transcription and determining the transcription start site, enhancers are responsible for increasing ("enhancing") transcription levels, and they are responsible for regulating cell- or tissue-specific transcription (i.e. the transcription responsible for differentiation). There are some other basic differences between enhancers and promoters, which are outlined in the module on transcription.

Enhancers function by being recognized and bound to by transcription factors. These are not the basal-type transcription factors (such as TFIID) discussed elsewhere. These are specialized transcription factors, of which there are very many types (all are proteins). A very large number of enhancer elements has been identified and characterized, and each different enhancer has its own transcription factor that it binds to.

TISSUE-SPECIFIC GENE EXPRESSION

How is a gene turned on in one cell type, and not in another? It depends primarily on whether the transcription factor for the gene's enhancer is active or not in a cell. To illustrate this, let's consider two different genes - one is regulated by enhancer A, which is recognized by transcription factor A, and the other gene is regulated by enhancer B, which is recognized by transcription factor B. In one cell type (let's say muscle, for the sake of argument), transcription factor A might be active whereas transcription factor B might not. In such a cell, only the first gene would be transcribed:

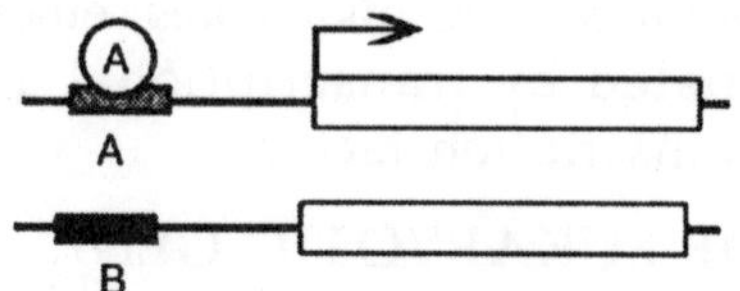

In another cell type (e.g. epidermis), transcription factor B would be active, and transcription factor A would not. In this cell, only the second gene would be transcribed:

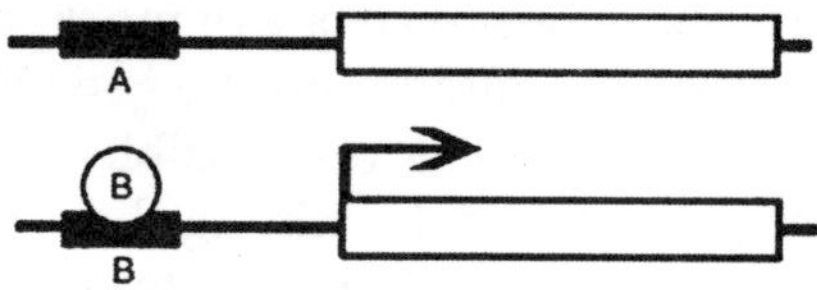

So how is it determined which transcription factors are active in which cell types? The answer to this question is very complicated, and not completely understood. However, there are a number of mechanisms by which transcription factors can be regulated.

Often, the presence or absence of a transcription factor in a cell is the determining step. If the factor is present, the gene is transcribed. If not, then the gene is not transcribed. The presence of a transcription factor, of course, depends upon the activity of the gene encoding that transcription factor, which forces us to ask how the gene encoding the transcription factor is regulated, which pretty much brings us back to where we began.

Transcription factors can be activated by environmental signals. For example, virtually all organisms have a set of genes called heat shock genes that encode proteins that help the organism survive heat stress. These genes are activated under conditions of heat stress, under the control of a transcription factor called heat shock transcription factor. This factor is always present, but is only activated when greatly increased temperatures are detected.

Transcription factors can be activated by signals from other cells in the same organism. Such signals include hormones and growth factors. Hormones must bind to a specific receptor on the target cell, and the receptor mediates the cellular effects of the signal. There are two basic mechanisms used, one for steroid hormones, and one for peptide hormones:

- Steroid hormones are lipid (actually cholesterol) derivatives, such as testosterone and progesterone. These hormones can cross the cytoplasmic membrane into a cell, where they bind to their specific receptor. Steroid receptors are transcription factors, and when they bind to their ligand, they become activated and initiate transcription of a specific set of genes.
- Peptide hormones cannot cross the cytoplasmic membrane, and so must bind to a receptor on the cell surface. When bound to its ligand, these receptors initiate a series of biochemical reactions inside the cell, with the ultimate result being the activation of a transcription factor (often by phosphorylation of the transcription factor), which initiates transcription of a specific set of genes.

DEVELOPMENTAL GENETICS

The primary function of the majority of genes in eukaryotic organisms is to coordinate the development of embryos. This module takes a look at some of the basic principles underlying the role of genes in embryonic development.

In most of the other modules of this course, the effects of genes are examined with regard to their effect on the phenotype of the juvenile or adult organism. In reality, however, the vast majority of these phenotypes are established during embryonic development, because most genes function to generate the pattern (the 'shape') of the developing embryo. Loss of function of a particular gene results in an abnormality in development, which manifests itself as a phenotype in the adult. In this sense, virtually all of eukaryotic genetics is developmental genetics, even though we haven't considered it as such. The field of developmental biology is concerned with the function of genes in embryogenesis. The central question of developmental biology is the following:

How does a single cell, the fertilized egg (or zygote), manage to produce an extremely complex adult organism, which is composed not only of trillions of cells, but of thousands of different types of cells (such as nerve cells, muscle cells, etc.)?

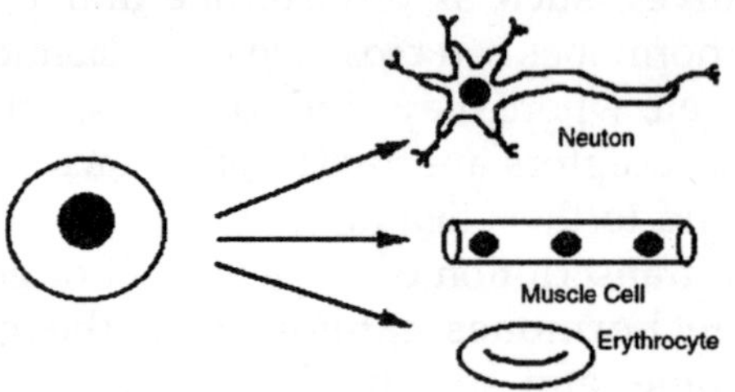

There is a simple answer to this question, and that is that cells differentiate. In other words, different cells become specialized to carry out different functions by following different developmental pathways. But how does this occur? How do cells know what pathway to follow?

MOSAIC THEORY OF DEVELOPMENT

One early theory that attempted to explain the process of differentiation was the mosaic theory, as outlined by Wilhelm Roux and August Weismann. This theory proposed that there were determinants that specified the various differentiation pathways. The theory stated that these determinants would all be present in the zygote, but as cell division occurred after fertilization, the determinants would be unequally inherited by the offspring cells. This is known as qualitative cell division. The set of determinants would thus be divided up until each cell contained only one type of determinant, and that determinant would determine the fate of that cell.

When Mendel's work was rediscovered in the early 1900's and the concept of the 'gene' was developed, it was believed that the genes were the determinants that were divided up during cell division. This idea is an appealing one because of its simplicity, but the theory is impeded by one small problem: *IT IS WRONG.*

Evidence was accumulated over sixty years that disproved the notion that genes are divided up in development. The final nail in the coffin of this theory came from nuclear transplantation studies done using amphibians in the early 1960's. In these studies, nuclei were isolated from tadpole intestinal cells (differentiated cells) and injected into eggs that had their own haploid nuclei removed. A small percentage of the injected eggs developed into completely normal adult frogs! These frogs had developed using only the genetic information found in a tadpole intestinal nucleus. Therefore, a tadpole intestinal nucleus must contain all of the genetic information necessary to allow the differentiation of every cell type in an adult frog. If the mosaic theory were correct, this would not be the case; a tadpole intestinal nucleus would have only the genetic information necessary to cause the differentiation of an intestinal cell.

Note: The frogs produced by this procedure would be genetically identical to the frog from which the intestinal cell nucleus was obtained. In other words, they would be clones. In fact, these experiments comprised the earliest successful cloning of vertebrate organisms. The recent cloning of Dolly the sheep was done for the same reasons as outlined above: demonstrating that differentiated mammalian nuclei (from mammary gland in this case) have all of the genetic information necessary to drive the development of a normal adult organism.

DIFFERENTIAL GENE EXPRESSION

If all nuclei in an organism contain the same genetic information, then how does differentiation occur? The explanation for this is found in the Theory of Differential Gene Expression. This theory states that differentiation occurs as a result of expression in a particular cell of only a subset of the total genes present. For example, if a cell expresses only the set of genes that causes muscle differentiation, then that cell will differentiate into a muscle cell. Alternatively, if the cell expresses only the set of genes that causes spleen cell differentiation, then the cell will differentiate into a spleen cell.

This is a fairly simple concept, but it hasn't really answered the question. The question has now become: how do cells activate only a certain set of genes (inactivating all other genes), and how do they know which set of genes to activate? The answer to this question is fairly straightforward as well: the set of genes activated in a cell is dependent on the set of transcription factors found in the cell. Using the muscle cell example from above, the cell activates the muscle-specific genes because it contains transcription factors that specifically activate the muscle-specific genes.

If we think about this for a minute, you'll realise that we still haven't answered the question; we've only changed it again. Now the question becomes: how did the cell come to contain only that specific set of transcription factors? Well, those transcription factors are encoded by genes, and those genes are regulated by other specific transcription factors. Those transcription factors are in turn encoded by other genes, which are regulated by still other transcription factors, etc.

As we can see, there is a hierarchy of genes within each cell. Genes are expressed that encode transcription factors, which activate other genes that encode transcription factors, which activate other genes that bring about differentiation along a specific pathway. How many levels are there to the hierarchy? From what we've seen so far, the hierarchy seems to go on forever. Each set of transcription factors is encoded by genes that are activated by yet another set of transcription factors. It must end somewhere, but where?

MASTER CONTROL GENES

In one sense, it ends with master control genes. A master control gene is the first gene activated in a hierarchy that leads to differentiation along a particular pathway. Master control genes encode the first transcription factor in a hierarchy; a master control gene product activates the next set of genes that encodes the next set of transcription factors, and the cascade of gene expression has been set in motion.

Let's look again at the muscle cell example considered earlier. The master control 'gene' for muscle development is actually a family of genes called the MyoD gene family. These genes encode HLH-type transcription factors. How do we know these are the master control genes for muscle? Master control genes have a particular property.

Because they initiate muscle differentiation, if the MyoD genes are activated in other cell types, they cause those cells to transdifferentiate into muscle. For example, hepatocytes (liver cells) or adipocytes (fat cells) that are caused to express the MyoD genes (using tricks of molecular biology) will change from their normal phenotype into muscle cells.

So how are the master control genes regulated? There are two basic ways that this can happen. One way is through the process of induction. Induction occurs when once cell sends a signal to another cell, telling it to differentiate a certain way. The signal, usually a diffusable protein, causes the recipient cell to activate the appropriate master control gene product. This is how muscle differentiation occurs. Signals from other cells cause the MyoD genes to become active in the cells receiving the signal, and muscle differentiation is initiated.

FUNCTIONS OF RIBOSOMES

Ribosomes are cytoplasmic granules composed of RNA and protein, at which protein synthesis takes place. They were first observed by Palade in the electron microscope as dense particles or granules. Upon isolation, they were shown to contain approximately equal amounts of RNA and protein Label. To function actively in protein synthesis, they must be bound into complete ribosomes.

We know that ribosomes are but one of the required components necessary for the synthesis of protein. The others are messenger RNA, which carries the genetic message; soluble RNA, which carries amino acids to be synthesized; and guanosine triphosphate, which is the source of energy. A number of ribosomes may be attached to the same messenger, each manufacturing its own chain of polypeptides, called a polysome. Ribosomes are also found in the mitochondria and chloroplasts of eukaryotic cells.

They are always smaller than the 80S cytoplasmic ribosomes, and are comparable to prokaryotic ribosomes in both size and sensitivity to antibiotics, although the sedimentation values vary somewhat in different phyla. Prokaryotic and eukaryotic ribosomes do not differ in any fundamental way; both preform the same functions by the same set of chemical reactions. The genetic code is the same in all living organism, and it has been demonstrated that eukaryotic ribosomes are able to translate bacterial mRNAs correctly. Eukaryotic ribosomes are much larger that prokaryotic ones and most of their proteins are different.

Antibiotics such as chloramphenicol inhibit bacterial but not eukaryotic ribosomes. Protein synthesis by eukaryotic ribosomes in inhibited by cycloheximide. Mitochondrial and chloroplast ribosomes resemble those in bacteria. They are inhibited by chloramphenicol, and hybrid ribosomes containing one bacterial and one chloroplast ribosome subunit, for example, are fully active in protein synthesis. Hybrid eukaryotic ribosomes containing subunits from both plants and mammals are also active in protein synthesis, but are inactive if one of the subunits is derived from bacteria.

Some structural resemblance must exist, however, since reconstruction experiments have shown that two proteins from the large subunit of E. coli can replace the homologous proteins in mammalian ribosomes. In summary, there is little structural but considerably functional homology between prokaryotic and eukaryotic ribosomes. Cells devote considerable effort to the production of these essential organells. For example, an E. coli cell contains approximately 15,000 ribosomes, each one with a molecular weight of about three million daltons. Ribosomes therefore represent twenty-five percent of the total mass of these bacterial cellsLabel.

STRUCTURE

Ribosomes are tiny particles, about 200 A. It is composed of both proteins and RNA; in fact it has approximately 37 - 62% RNA, and rest are made up of proteins Label. The RNA present in ribosomes are obviously called ribosomal RNA, and they are produced in the nucleolus, which is a prominent globular structure in the nucleus. Thus, the proteins are gene products of themselves, and one ribosome is made up of dozens of genes. The ribosomes fall into two categories: Those that are free to roam in the cytoplasm, and those that are bound to gigantic, cobwebby organelles made up of membranes, called the endoplasmic reticulum; thus, causing a rough surface. Although, the two kinds of ribosomes play similar roles in translating mRNA to produce proteins, they are very distinct in where its product is located.

The ribosomes in the cytoplasm allows its protein to roam about freely, while the bound ribosomes transfer their functional protein into the endoplasmic reticulum. In addition, ribosomes are also located within the mitochondria, and the chloroplast, but are only few in content.

This spherical particle of 23nm, is composed of two subunits; a large and small Label. In Eukaryotes, the co-efficient of ribosomes are 80s, of which is divided into 60s for the large, and 40s for the small subunit. The 60s contain 28s rRNA, with a small fragment that is attached noncovalently and can be released upon heating; a 5.8s, and a very small - 120 nucleated of 5sRNA. Whereas, the 40s subunit has only a single 18s rRNA Label. In prokaryotes, however, the large and small subunits are split into 50s and 30s, making a total of 70s respectively.

The 50s has two types of rRNA - a 23s and a 5s. It also has 32 different proteins. On the other hand, the 30s contains a single 16s rRNA, 21 different types of proteins. To help better understand what the s stands for in rRNA, let us use the prokaryotes as an example. The 50s and 30s refers to the sedimentation coefficient of the two subunits. This coefficient is a measure of the speed with which the particles sediment through a solution when spun in an ultra centrifuge.

Thus, the particles with larger coefficient would centrifuge and settle much faster since it is has more mass than the particle with the smaller coefficient. 50s + 30s → 70s Note that the two subunits above make up the entire ribosomal molecule which is 70s. The reason the coefficients do not add up is because they are not proportional to the particle weight. During protein synthesis, ribosomes line up along the mRNA and form a polysome, also called the polyribosome. The mRNA is aligned in the gap between the 2 ribosomal subunits. It is possible that the nascent peptide chain grows through a channel or groove in the large ribosomal subunit. This is predicted to be the case since ribosomes protect a segment of 30-40 amino acids from degradation.

Speaking of amino acids, up to 30 ribosomes can attach on one strand of mRNA to form amino acid chains thus leading to protein formation. Ribosomes act as the backbone for many molecules during translation. It provides room for many structures to situate itself thus enhancing protein synthesis. For example, mRNA inserts itself between the two subunits; the peptidyl transferase complex - the enzyme that allows for the tRNA to break apart from the amino acid on P-site; this enzyme lays across the molecule, between the subunits. It contains the P and the A-site for tRNA binding.

Last but not least, the ribosome molecule allows the growing polypeptide chain, to emerge from the back of the structure, thus it is situated perpendicular to the mRNA chain. Ribosomes have a tertiary structure. Ribosomes make up a large part of cells in many species, which leads to protein manufacturing. For example, in E.Coli (bacteria), they make up about 1/4 of the total cell mass.

They are intensely basiphilic (having high affinity for bases). Due to its complex structures, with many proteins and different kinds of RNA, researchers have found it very difficult to study the macro molecular structure of ribosomes, especially for the fact it is quite impossible to observe its crystal using an x-ray diffraction. Thus, scientists have been forced to use other means of study to map the proteins and RNA components in ribosome. Some of these are the cross-linking, immunoelectron microscopy, and low-angle neutron scattering methods.

The cross-linking shows the protein arrangement and the types of bonds it forms within itself. The neutron scattering experiments forms horizontal lines that show the entire structure of ribosome, with its two subunits, and shows where the proteins are arranged in the molecule. The empty regions around the proteins is where the rRNA is located. The immunoelectron microscopy, shows the proposed location of the 16s rRNA molecule of the small subunit, in prokaryotes.

FUNCTION

The ribosomes plays a very important role in protein synthesis, which is the process by which proteins are made from individual amino acids. Without the ribosomes the message would not be read, thus proteins could not be produced. Therefore, ribosomes play a very important role in role in protein synthesis. The primary agent in the process of translating the mRNA into a specific amino acid chain is the ribosome, which consists of two subunits. These subunits are made up of a third and extremely abundant type of RNA, ribosomal RNA (rRNA), and together contain up to eighty-two specific proteins assembled in a precise sequence Label. The ribosomes constituents must be put together in an extremely precise position and sequence.

This assembled ribosome displays a series of small groves, tunnels, and platforms, where the action of protein synthesis occurs Label. There are the active sites, each dedicated to one of the tasks required for translation of mRNA into protein. Proteins being synthesized for export out of the cell, are made by ribosomes attached to the rough endoplasmic reticulum. In contrast, proteins for use by the cell are generally made in the cytoplasm by free ribosomes. Several of these free ribosomes may attach to a single mRNA molecule, giving rise to the polyribosome or polysome Label. Protein synthesis takes place on polyribosomes (or polysomes) where 80S ribosomes associate with an mRNA coding for a given protein. The number of ribosomes associated in the polysomal chains depends on the size of the mRNA.

This is also associated with the size of the protein that is being synthesized. Outside the polyribosome, the ribosomes are dissociated and form a pool of free subunits. Transfer RNAs are also bound to the ribosome. There are quite a few factors involved in the formation of the initiation complex. These include: GTP, methionine tRNA, an initiation codon in mRNA,

80S ribosomes, and three protein factors Label. The process of protein synthesis begins with the capture of the tRNA, which is carrying an amino acid, by an initiation factor.

This binds to a small ribosomal subunit, which occupies one of the active sites in the ribosomes, the P (protein) site. This initiation complex recognized and binds to the 5' end of an mRNA molecule and slides down to the initiation codon, which is always an AUG sequence of amino acids. The large subunit of the ribosome now joins the complex. A second tRNA is now brought into the ribosome by the elongation factor. If the anticodon of the tRNA pairs with the next codon of the message, the tRNA occupies the A (acceptor) site on the ribosome. This positions the second amino acid adjacent to the initiation methionine.

Then an enzyme, peptidyl transferase, which is part of the large ribosomal subunit mediates the separation of the first amino acid from its tRNA and the formation of a peptide bond between the initial methionine and the amino acid is formed. The P site is now occupied by an uncharged tRNA molecule label. The ribosome will now move down the mRNA by one codon, a process known as translocation. This movement shifts the growing polypeptide chain to the P position, and results in an empty A site, where a new charged tRNA can enter and pair, by forming a hydrogen bond between the codon and the anticodon. This holds the tRNA into place long enough for an even more stable binding to occur Label.

The uncharged tRNA that previously occupied the P site is booted out of the ribosome and will be recharged and recycled by the cell. The energy needed for this process is supplied by the hydrolysis of guanosine triphosphate (GTP). The process then continues along the length of the mRNA, until the first stop codon is encountered. At that point the action of a termination factor releases the completed protein from the last tRNA and the ribosome dissociates into its component parts. Another function of the ribosomes occurs in the relation to the neuron and axons.

The cell body of a typical large neuron contains vast

numbers of ribosomes. Although dendrites often contain some ribosomes, there are no ribosomes in the axon, and its protein must therefore be provided by the many ribosomes in the cell body.

STRUCTURE LINKAGE

Antibiotics are drugs produced by bacteria and fungi. These molecules function as drugs used in the chemotherapy of infectious disease, and follow the principle of selective toxicity. Selective toxicity follows the principle of using drugs that kill the harmful microorganism without damaging the host. As a result of its toxicity, antibiotics can affect the ribosomal structure, inhibiting protein synthesis. For instance, let us take the 70s ribosome of prokaryotes; antibiotics can target this structure and can adverse the effects on the cells of the host.

Among the antibiotics that interfere with protein synthesis are chloramphenicol, erythromyocin, streptomycin, and the tetracycline. This paper will be focused on these four antibioticsand its role played in the effect of the ribosome structure, thus leading to the change in protein synthesis. For instance, the chloramphenicol reacts with the 50s structure of the 70s prokaryote ribosome, by inhibiting the formation of the peptide bonds in the growing polypetide chain. Erythromyocin, the second antibiotic, also reacts with the same structure as chloromphenicol.

However, it has a very narrow range of activity, since it affects mostly the gram-positive bacteria. The other two antibiotics attract the 30s structure of the 70s prokaryotic ribosome. The tetracycline interferes with the attachment of the tRNA, which carries the amino acids, to the ribosome, thus preventing the addition of amino acids to the growing polypeptide chain. One unique aspect of tetracycline is that it cannot penetrate well into the mammalian cells, therefore, it does not interfere with the mammalian ribosomes. Aminoglycoside antibiotics, a type of streptomycin, changes the shape of the 30s structure of the 70s prokaryotic ribosome, thus interfering with the initial stage of protein synthesis. This in turn, causes the misreading of the genetic code on the mRNA.

FUNCTION LINKAGE

RNA polymerase is the enzyme that directs transcription, which is the process by which the mRNA copy of a gene is synthesized. Transcription follows the same rules of base pairing as DNA replication. This base pattern ensure that an RNA transcript is a faithful copy of the gene. There are three stages of transcription: initiation, elongation, and termination. During initiation, the enzyme recognizes a promoter region, which lies upstream from the gene. The polymerase binds tightly to the promoter and causes localized melting, or separation of the two DNA strands within the promoter. Then the polymerase starts building the RNA chain. Ribonucleoside triphosphates such as ATP, GTP, CTP and UTP are the building blocks the polymerase uses for this job Label.

After the first nucleotide is in place, the polymerase binds the second nucleotide, joining it to the first. This forms the initial phosphodiester bond in the RNA chain. The second stage is elongation, where the RNA polymerase directs the sequential binding of ribonucleotides to the RNA chain. As it does this, it moves along the DNA template and the melted DNA moves with it. This melted region exposes the bases of the template DNA one-by-one so that they can pair with the bases of the incoming ribonucleotides.

As soon as the transcription machinery passes, the two DNA strands wind around each other again, reforming the double helix. Only enough separation will occur so that the polymerase can read the DNA template Label. The final stage is termination, which allows the termination of transcription. These work in conjunction with RNA polymerase, and is sometimes aided by another protein, to loosen the association between RNA product and DNA template. So the RNA dissociates from the RNA polymerase and DNA, thus terminating transcription. Transcription is very important in that it is the only step in expression of the genes for rRNAs.

It is also important to ribosomes, because it sets up the

RNA in a 5' to 3' sequence, which allows the ribosomes to read the message 5' to 3'. Without transcription of a gene, the ribosome would not be able to translate an mRNA, thus not allowing it to be separated into its component parts. If this does not occur, then translation will also not occur, thereby affecting the function and role of ribosomes in translation.

Regulation linkage: Ribosomes are used by all living cells to synthesize proteins. This synthesis can be inhibited by antibiotics, which can have an effect on the organism. Some antibiotics target specific subunits of the ribosome or may target the entire ribosome completely. Streptomycin, for example, can target the 70s ribosome in some prokaryotes and cause adverse effects on the cell of the host.

An antibiotic that affects the 30s ribosome is tetracycline. It can prevent the addition of amino acids to the growing polypeptide chain by interfering with the attachment of the tRNA onto the ribosome. The 50s ribosome may also be targeted by erythromycin, which blocks the translocation reaction on ribosomes. Other antibiotics that interfere with the ribosome to synthesize proteins include chloramphenicol, rifamycin, puromycin, cycloheximide, and anisomycin.

MESSENGER RNA

Messenger ribonucleic acid (mRNA) is a molecule of RNA encoding a chemical "blueprint" for a protein product. mRNA is transcribed from a DNA template, and carries coding information to the sites of protein synthesis: the ribosomes. Here, the nucleic acid polymer is translated into a polymer of amino acids: a protein. In mRNA as in DNA, genetic information is encoded in the sequence of four nucleotides arranged into codons of three bases each.

Each codon encodes for a specific amino acid, except the stop codons that terminate protein synthesis. This process requires two other types of RNA: transfer RNA (tRNA) mediates recognition of the codon and provides the corresponding amino acid, while ribosomal RNA (rRNA) is the central component of the ribosome's protein manufacturing machinery.

MRNA

The brief existence of an mRNA molecule begins with transcription and ultimately ends in degradation. During its life, an mRNA molecule may also be processed, edited, and transported prior to translation. Eukaryotic mRNA molecules often require extensive processing and transport, while prokaryotic molecules do not.

Transcription

During transcription, RNA polymerase makes a copy of a gene from the DNA to mRNA as needed. This process is similar in eukaryotes and prokaryotes. One notable difference, however, is that eukaryotic RNA polymerase associates with mRNA processing enzymes during transcription so that processing can proceed quickly after the start of transcription. The short-lived, unprocessed or partially processed, product is termed *pre-mRNA*; once completely processed, it is termed *mature mRNA*.

Eukaryotic pre-mRNA Processing

Processing of mRNA differs greatly among eukaryotes, bacteria and archea. Non-eukaryotic mRNA is essentially mature upon transcription and requires no processing, except in rare cases. Eukaryotic pre-mRNA, however, requires extensive processing.

5′ Cap Addition

A *5′ cap* (also termed an RNA cap, an RNA 7-methylguanosine cap or an RNA m^7G cap) is a modified guanine nucleotide that has been added to the "front" or 5' end of a eukaryotic messenger RNA shortly after the start of transcription. The 5' cap consists of a terminal 7-methylguanosine residue which is linked through a 5'-5'-triphosphate bond to the first transcribed nucleotide. Its presence is critical for recognition by the ribosome and protection from RNases.

Cap addition is coupled to transcription, and occurs co-transcriptionally, such that each influences the other. Shortly after the start of transcription, the 5' end of the mRNA being synthesized is bound by a cap-synthesizing complex associated with RNA polymerase. This enzymatic complex catalyzes the chemical reactions that are required for mRNA capping. Synthesis proceeds as a multi-step biochemical reaction.

Splicing

Splicing is the process by which pre-mRNA is modified to remove certain stretches of non-coding sequences called introns; the stretches that remain include protein-coding sequences and are called exons. Sometimes pre-mRNA messages may be spliced in several different ways, allowing a single gene to encode multiple proteins. This process is called alternative splicing. Splicing is usually performed by an RNA-protein complex called the spliceosome, but some RNA molecules are also capable of catalyzing their own splicing.

Editing

In some instances, an mRNA will be edited, changing the nucleotide composition of that mRNA. An example in humans is the apolipoprotein B mRNA, which is edited in some tissues, but not others. The editing creates an early stop codon, which upon translation, produces a shorter protein.

POLYADENYLATION

Polyadenylation is the covalent linkage of a polyadenylyl moiety to a messenger RNA molecule. In eukaryotic organisms, most messenger RNA (mRNA) molecules are polyadenylated at the 3' end. The poly(A) tail and the protein bound to it aid in protecting mRNA from degradation by exonucleases. Polyadenylation is also important for transcription termination, export of the mRNA from the nucleus, and translation. mRNA can also be polyadenylated in prokaryotic organisms, where poly(A) tails act to facilitate, rather than impede, exonucleolytic degradation.

Polyadenylation occurs during and immediately after transcription of DNA into RNA. After transcription has been terminated, the mRNA chain is cleaved through the action of an endonuclease complex associated with RNA polymerase. The cleavage site is characterized by the presence of the base sequence AAUAAA near the cleavage site. After the mRNA has been cleaved, 80 to 250 adenosine residues are added to the free 3' end at the cleavage site. This reaction is catalyzed by polyadenylate polymerase. Just as in alternative splicing, there can be more than one polyadenylation variant of a mRNA.

TRANSPORT

Another difference between eukaryotes and prokaryotes is mRNA transport. Because eukaryotic transcription and translation is compartmentally separated, eukaryotic mRNAs must be exported from the nucleus to the cytoplasm. Mature mRNAs are recognized by their processed modifications and then exported through the nuclear pore.

TRANSLATION

Because prokaryotic mRNA does not need to be processed or transported, translation by the ribosome can begin immediately after the end of transcription. Therefore, it can be said that prokaryotic translation is *coupled* to transcription and occurs *co-transcriptionally*.

Eukaryotic mRNA that has been processed and transported to the cytoplasm (i.e. mature mRNA) can then be translated by the ribosome. Translation may occur at ribosomes free-floating in the cytoplasm, or directed to the endoplasmic reticulum by the signal recognition particle. Therefore, unlike prokaryotes, eukaryotic translation *is not* directly coupled to transcription.

DEGRADATION

After a certain amount of time, the message is degraded by RNases. The limited lifetime of mRNA enables a cell to alter protein synthesis rapidly in response to its changing needs. Different mRNAs within the same cell have distinct lifetimes (stabilities). In bacterial cells, individual mRNAs can survive from seconds to more than an hour; in mammalian cells, mRNA lifetimes range from several minutes to days. The greater the stability of an mRNA, the more protein may be produced from that mRNA.

The presence of AU-rich elements in some mammalian mRNAs tends to destabilize those transcripts through the action of cellular proteins that bind these motifs. Rapid mRNA degradation via AU-rich elements is a critical mechanism for preventing the overproduction of potent cytokines such as tumor necrosis factor (TNF) and granulocyte-macrophage colony stimulating factor (GM-CSF). Base pairing with a small interfering RNA (siRNA) or microRNA (miRNA) can also accelerate mRNA degradation.

mRNA STRUCTURE

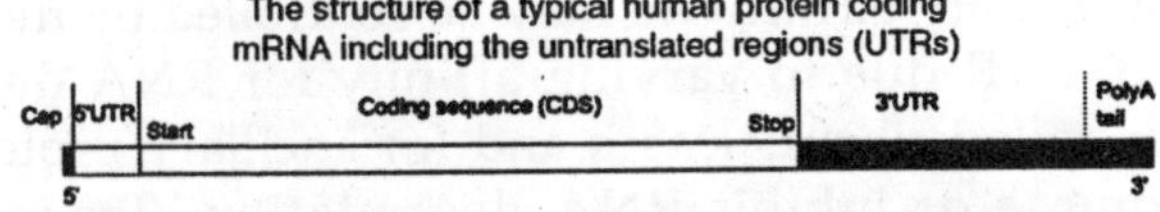

Fig. The Structure of a Mature Eukaryotic mRNA. A Fully Processed mRNA Includes a 5' cap, 5' UTR, Coding Region, 3' UTR, and Poly(A) tail

5' Cap

The *5' cap* is a modified guanine nucleotide added to the "front" (5' end) of the pre-mRNA using a 5',5-Triphosphate linkage. This modification is critical for recognition and proper attachment of mRNA to the ribosome, as well as protection from 5' exonucleases. It may also be important for other essential processes, such as splicing and transport.

Coding Regions

Coding regions are composed of codons, which are decoded and translated into one (mostly eukaryotes) or several (mostly prokaryotes) proteins by the ribosome. Coding regions begin with the start codon and end with the one of three possible stop codons. In addition to protein-coding, portions of coding regions may also serve as regulatory sequences in the pre-mRNA as exonic splicing enhancers or exonic splicing silencers. Start codons are indicated by a AUG triplet. Stop codons are indicated by a UAA, UAG, or UGA.

Untranslated Regions

Untranslated regions (UTRs) are sections of the mRNA before the start codon and after the stop codon that are not translated, termed the five prime untranslated region (5' UTR) and three prime untranslated region (3' UTR), respectively. These regions are transcribed with the coding region and thus are exonic as they are present in the mature mRNA. Several roles in gene expression have been attributed to the untranslated regions, including mRNA stability, mRNA localization, and translational efficiency. The ability of a UTR to perform these functions depends on the sequence of the UTR and can differ between mRNAs.

The stability of mRNAs may be controlled by the 5' UTR and/or 3' UTR due to varying affinity for RNA degrading enzymes called ribonucleases and for ancillary proteins that can promote or inhibit RNA degradation. Translational efficiency, including sometimes the complete inhibition of translation, can be controlled by UTRs. Proteins that bind to either the 3' or 5' UTR may affect translation by influencing the ribosome's ability to bind to the mRNA.

MicroRNAs bound to the 3' UTR also may affect translational efficiency or mRNA stability. Cytoplasmic localization of mRNA is thought to be a function of the 3' UTR. Proteins that are needed in a particular region of the cell can actually be translated there; in such a case, the 3' UTR may contain sequences that allow the transcript to be localized to this region for translation.

Some of the elements contained in untranslated regions form a characteristic secondary structure when transcribed into RNA. These structural mRNA elements are involved in regulating the mRNA. Some, such as the SECIS element, are targets for proteins to bind. One class of mRNA element, the riboswitches, directly bind small molecules, changing their fold to modify levels of transcription or translation. In these cases, the mRNA regulates itself.

3' Poly(A) Tail

The 3' poly(A) tail is a long sequence of adenine nucleotides (often several hundred) added to the "tail" or 3' end of the pre-mRNA through the action of an enzyme, polyadenylate polymerase. In higher eukaryotes, the poly(A) tail is added onto transcripts that contain a specific sequence, the AAUAAA signal. The importance of the AAUAAA signal is demonstrated by a mutation in the human alpha 2-globin gene that changes the original sequence AATAAA into AATAAG, which can lead to hemoglobin deficiencies.

Monocistronic Versus Polycistronic Mrna

An mRNA molecule is said to be monocistronic when it contains the genetic information to translate only a single protein. This is the case for most of the eukaryotic mRNAs. On the other hand, polycistronic mRNA carries the information of several proteins, which are translated into several proteins. Most of the mRNA found in bacteria and archea are polycistronic. Dicistronic is the term used to describe a mRNA that encodes only two proteins.

TRANSFER RNA

Transfer RNA (abbreviated tRNA) is a small RNA (usually about 74-95 nucleotides) that transfers a specific amino acid to a growing polypeptide chain at the ribosomal site of protein synthesis during translation. It has a 3' terminal site for amino acid attachment. This covalent linkage is catalyzed by an aminoacyl tRNA synthetase. It also contains a three base region called the anticodon that can base pair to the corresponding three base codon region on mRNA. Each type of tRNA molecule can be attached to only one type of amino acid, but because the genetic code contains multiple codons that specify the same amino acid, tRNA molecules bearing different anticodons may also carry the same amino acid.

STRUCTURE

tRNA has primary structure, secondary structure (usually visualized as the *cloverleaf structure*), and tertiary structure (all tRNAs have a similar L-shaped 3D structure that allows them to fit into the P and A sites of the ribosome).

- The 5'-terminal phosphate group.
- The acceptor stem is a 7-bp stem made by the base pairing of the 5'-terminal nucleotide with the 3'-terminal nucleotide (which contains the CCA 3'-terminal group used to attach the amino acid). The acceptor stem may contain non-Watson-Crick base pairs.
- The CCA tail is a CCA sequence at the 3' end of the tRNA molecule. This sequence is important for the recognition of tRNA by enzymes critical in translation. In prokaryotes, the CCA sequence is transcribed. In eukaryotes, the CCA sequence is added during processing and therefore does not appear in the tRNA gene.
- The D arm is a 4 bp stem ending in a loop that often contains dihydrouridine.
- The anticodon arm is a 5-bp stem whose loop contains the anticodon.
- The T arm is a 5 bp stem containing the sequence TØC where Ø is a pseudouridine.
- Bases that have been modified, especially by methylation, occur in several positions outside the anticodon. The first anticodon base is sometimes modified to inosine (derived from adenine) or pseudouridine (derived from uracil).

ANTICODON

An anticodon is a unit made up of three nucleotides that correspond to the three bases of the codon on the mRNA. Each tRNA contains a specific anticodon triplet sequence that can base-pair to one or more codons for an amino acid. For example, one codon for lysine is AAA; the anticodon of a lysine tRNA might be UUU.

Some anticodons can pair with more than one codon due to a phenomenon known as wobble base pairing. Frequently, the first nucleotide of the anticodon is one of two not found on mRNA: inosine and pseudouridine, which can hydrogen bond to more than one base in the corresponding codon position. In the genetic code, it is common for a single amino acid to be specified by all four third-position possibilities; for example, the amino acid glycine is coded for by the codon sequences GGU, GGC, GGA, and GGG.

To provide a one-to-one correspondence between tRNA molecules and codons that specify amino acids, 61 tRNA molecules would be required per cell. However, many cells contain fewer than 61 types of tRNAs because the wobble base is capable of binding to several, though not necessarily all, of the codons that specify a particular amino acid.

AMINOACYLATION

Aminoacylation is the process of adding an aminoacyl group to a compound. It produces tRNA molecules with their CCA 3' ends covalently linked to an amino acid. Each tRNA is aminoacylated (or *charged*) with a specific amino acid by an aminoacyl tRNA synthetase. There is normally a single aminoacyl tRNA synthetase for each amino acid, despite the fact that there can be more than one tRNA, and more than one anticodon, for an amino acid. Recognition of the appropriate tRNA by the synthetases is not mediated solely by the anticodon, and the acceptor stem often plays a prominent role.

Reaction:

- Amino acid + ATP → aminoacyl-AMP + PPi
- Aminoacyl-AMP + tRNA → aminoacyl-tRNA + AMP

BINDING TO RIBOSOME

The ribosome has three binding sites for tRNA molecules: the A, P and E sites. During translation the A site binds an incoming aminoacyl-tRNA as directed by the codon currently occupying this site. This codon specifies the next amino acid to be added to the growing peptide chain. The A site only works after the first aminoacyl-tRNA has attached to the P site. The P-site codon is occupied by peptdyl-tRNA that is a tRNA with multiple amino acids attached as a long chain.

The P site is actually the first to bind to aminoacyl tRNA. This tRNA in the P site carries the chain of amino acids that has already been synthesized. The E site is occupied by the empty tRNA as it is about to exit the ribosome.

tRNA GENES

Organisms vary in the number of tRNA genes in their genome. The nematode worm *C. elegans,* a commonly used model organism in genetics studies, has 29,647 genes in its nuclear genome, of which 620 code for tRNA. The budding yeast *Saccharomyces cerevisiae* has 275 tRNA genes in its genome. In the human genome, which according to current estimates has about 27,161 genes in total, there are about 4,421 non-coding RNA genes, which include tRNA genes. There are 22 mitochondrial tRNA genes; 497 nuclear genes encoding cytoplasmic tRNA molecules and there are 324 tRNA-derived putative pseudogenes.

Cytoplasmic tRNA genes can be grouped into 49 families according to their anticodon features. These genes are found on all chromosomes, except 22 and Y chromosome. High clustering on 6p is observed (140 tRNA genes), as well on 1 chromosome. tRNA molecules are transcribed (in eukaryotic cells) by RNA polymerase III, unlike messenger RNA which is transcribed by RNA polymerase II. pre-tRNAs contain introns; in bacteria these self-splice, whereas in eukaryotes and archaea they are removed by tRNA splicing endonuclease.

The existence of tRNA was first hypothesized by Francis Crick, based on the assumption that there must exist an adapter molecule capable of mediating the translation of the RNA alphabet into the protein alphabet. Significant research on structure was conducted in the early 1960s by Alex Rich and Don Caspar, two researchers in Boston, the Jacques Fresco group in Princeton University and a United Kingdom group at King's College London.

A later publication reported the primary structure in 1965 by Robert W. Holley. The secondary and tertiary structures were derived from X-ray crystallography studies reported independently in 1974 by American and British research groups headed, respectively, by Alexander Rich and Aaron Klug.

MODIFIED BASES IN tRNA

Transfer RNA molecules (tRNAs) are typically about 75 nucleotides long and fold into stable tertiary structures not unlike polypeptides. While they have no independent chemical function, their structures are finely tuned to suit a number of steps in translation. They must be specifically recognized by the enzymes that attach amino acids to them, they must bind efficiently to the catalytic sites in the ribosome and they must form productive interactions with mRNA transcripts.

The focus of this exercise will be on the adaptation of tRNA structure that permits efficient and specific interactions with mRNA. The region of the tRNA molecule that interacts with the messenger is known as the anticodon. It is comprised of three nucleotides that form part of the anticodon loop. These three bases are free to hydrogen bond with other nucleotides, particularly a set of three base sequences (codons) on mRNA that correspond to the amino acid for which the tRNA molecule is specific. In principle, all three bases of the anticodon could interact via normal Watson- Crick base-pairing to form a three base pair duplex with a single codon sequence.

In fact, the first nucleotide of the anticodon, position 34 in tRNA(Phe), which pairs with the third base of the codon, is known as the "wobble" base. The degeneracy of the genetic code means that several codons can specify a single amino acid. To reduce the number of tRNAs necessary to read the codons, the wobble position is often intended to base pair with two or three different bases at the 3' end of the codon.

This reduces the specificity of interaction between the tRNA molecule and the transcript, and has the potential to lead to significant errors in translation if only two base pairs provide the thermodynamic stability to determine the interaction.

It appears that one approach taken by nature to enhance the specific base pairing interactions between tRNA and mRNA is to use modified bases at positions flanking the anticodon in the tRNA molecule. tRNA is notable for the number and variety of modified bases in its structure.

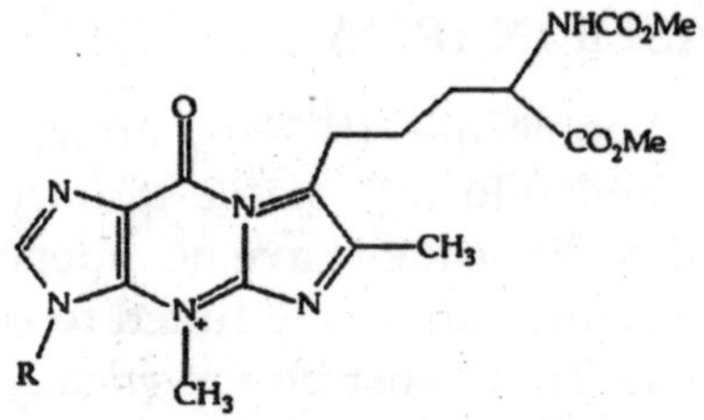

Fig. The Structure of Wybutosine

These bases are typically modified forms of the usual A, U, C and G's. Some are methylated analogues (for example 5-methyl uracil - otherwise known as thymine - is found in tRNA), and others are more dramatic, such as wybutosine. These modifications are post-transcriptional. They are made after the tRNA molecule is transcribed from its gene. The purpose of this exercise is to investigate how modifications made to tRNA(Phe) from yeast, including the creation of wybutosine adjacent to the anticodon, aid in the function of that molecule.

tRNA(Phe) has an anticodon sequence GAA, that binds to sequences UUC and UUU on the messenger (both codons for phenylalanine). Because of the wobble at position 3 of the codon, and the U:A base pairs at positions 1 and 2, the interaction between tRNA(Phe) and mRNA is weaker than most. Here, we will examine how this potential problem is overcome.

THE MODEL

Three models are required for this exercise, though a number of others will need to be created in the course of the work. The first comes from coordinates for tRNA(Phe) deposited to the PDB by Kim and co-workers.*(1)* 6tna.pdb is a slightly modified version of the actual PDB entry, with a few changes made for ease of manipulation in Midas.

The The second model (arna.pdb is a undecamer duplex of RNA, derived from idealized coordinates of RNA in the A conformation, containing the sequence UUC at the 5' end of one of the strands. brna.pdb is a decamer of duplex RNA, like rna.pdb, but in the B conformation.

RIBOSOMAL RNA

Ribosomal RNA (rRNA) is the central component of the ribosome, the protein manufacturing machinery of all living cells. The function of the rRNA is to provide a mechanism for decoding mRNA into amino acids and to interact with the tRNAs during translation by providing peptidyl transferase activity.

INSIDE THE RIBOSOME

The ribosome is composed of two subunits, named for how rapidly they sediment when subject to centrifugation. tRNA is sandwiched between the small and large subunits and the ribosome catalyzes the formation of a peptide bond between the 2 amino acids that are contained in the tRNA.

The ribosome also has 3 binding sites called A, P, and E.

- The A site in the ribosome binds to an aminoacyl-tRNA (a tRNA bound to an amino acid).
- The NH_2 group of the aminoacyl-tRNA which contains the new amino acid, attacks the carboxyl group of peptidyl-tRNA (contained within the P site) which contains the last amino acid of the growing chain called peptidyl transferase reaction.
- The tRNA that was holding on the last amino acid is moved to the E site, and what used to be the aminoacyl-tRNA is now the peptidyl-tRNA.

A single mRNA can be translated simultaneously by multiple ribosomes.

PROKARYOTES VS. EUKARYOTES

Both prokaryotic and eukaryotic can be broken down into two subunits (the S in 16S represents Svedberg units):

Type	Size	Large Subunit	Small Subunit
prokaryotic	70S	50S (5S, 23S)	30S (16S)
eukaryotic	80S	60S (5S, 5.8S, 28S)	40S (18S)

Note that the S units of the subunits cannot simply be added because they represent measures of sedimentation rate rather than of mass. The sedimentation rate of each subunit is affected by its shape, as well as by its mass.

Prokaryotes

In prokaryotes a small 30S ribosomal subunit contains the 16S rRNA. The large 50S ribosomal subunit contains two rRNA species (the 5S and 23S rRNAs). Bacterial 16S, 23S, and 5S rRNA genes are typically organized as a co-transcribed operon. There may be one or more copies of the operon dispersed in the genome (for example, *Escherichia coli* has seven).

Archaea contains either a single rDNA operon or multiple copies of the operon. The 3' end of the 16S rRNA (in a ribosome) binds to a sequence on the 5' end of mRNA called the Shine-Dalgarno sequence.

Eukaryotes

In contrast, eukaryotes generally have many copies of the rRNA genes organized in tandem repeats; in humans approximately 300–400 rDNA repeats are present in five clusters (on chromosomes 13, 14, 15, 21 and 22).

The 18S rRNA in most eukaryotes is in the small ribosomal subunit, and the large subunit contains three rRNA species (the 5S, 5.8S and 28S rRNAs).

Mammalian cells have 2 mitochondrial (12S and 16S) rRNA molecules and 4 types of cytoplasmic rRNA (28S, 5.8S, 5S (large ribosome subunit) and 18S (small subunit). 28S, 5.8S, and 18S rRNAs are encoded by a single transcription unit (45S) separated by 2 Internally transcribed spacer (ITS). The 45S rDNA organized into 5 clusters (each has 30-40 repeats) on chromosomes 13, 14, 15, 21, and 22. These are transcribed by RNA polymerase I. 5S occurs in tandem arrays (~200-300 true 5S genes and many dispersed pseudogenes), the largest one on the chromosome 1q41-42. 5S rRNA is transcribed by RNA polymerase III.

The tertiary structure of the small subunit ribosomal RNA (SSU rRNA) has been resolved by X-ray crystallography. The secondary structure of SSU rRNA contains 4 distinct domains — the 5', central, 3' major and 3' minor domains. A model of the secondary structure for the 5' domain (500-800 nucleotides) is shown.

Translation

Translation is the net effect of proteins being synthesized by ribosomes, from a copy (mRNA) of the DNA template in the nucleus. One of the components of the ribosome (16s rRNA) base pairs complementary to a sequence upstream of the start codon in mRNA.

IMPORTANCE OF rRNA

Ribosomal RNA characteristics are important in medicine and in evolution.

- rRNA is the target of several clinically relevant antibiotics: chloramphenicol, erythromycin, kasugamycin, micrococcin, paromomycin, ricin, sarcin, spectinomycin, streptomycin, and thiostrepton.
- rRNA is the most conserved (least variable) gene in all cells. For this reason, genes that encode the rRNA (rDNA) are sequenced to identify an organism's taxonomic group, calculate related groups, and estimate rates of species divergence. For this reason many thousands of rRNA sequences are known and stored in specialized databases such as RDP-II and the European SSU database.

SYNTHESIS OF RIBOSOMAL RNA - rRNA

The 70S ribosome of prokaryotes consists of a 308 subunit and a 50S subunit. The former contains 16S rRNA and the latter 238 and 58 rRNAs. In bacterial genes the sequences specifying 168, 238 and some times also5S RNA are arranged in a series mRNA is transcribed from DNA as a 30S unit, which bas been called p308. Experimental evidence indicates that the 308 unit has the 16S component at the 5' end and the 238 component at the 3' end, with spacer units between the two components. In some prokaryotes 5S rRNA is transcribed at or near the 3' end.

During processing the p30S transcriptional unit is cleaved within the spacer segment by RNase III into 25S and l8S segments. These are then reduced to p23S and pl6S segments respectively, also by RNase III. Secondary trimming of these intermediates yields the final size, 23S and 168, respectively.

The enzyme involved in secondary trimming is probably a ribonuclease, designated as RNase M or maturase.In E. coli trimming of the pl6s component involves removal of a total of about 200 nucleotides from both 3' and 5' sides. The final cleavages occur when the rRNAs are associated with structural proteins of ribosomes in the 'preribosomal particles'.

Some nucleoside modifications take place in the p30S component, while others occur only after cleavage. Methylaiions found in 23S rRNA occur early at the p30S stage, whereas modifications found in 168 rRNA take place at a late stage of processing, in some cases after association with specific ribosomal proteins.

In E. coli the precursors of 58 rRNA are only a few nucleotides larger than mature SS rRNA. In Bacillus there are two types of larger precursors. One, designated as p5A, is 180 nucleotides long, while the other, P5B' is 140-150 nucleotides long. Both are apparently transcribed on different 5S genes.

Cleavage of p5A by the enzyme RNase M5 splits it into three components, a 5' terminal piece of about 20 nucleotides, mature 5S rRNA of 118 nucleotides and a 3' terminal piece of about 40 nucleotides. The 5' and 3' terminal pieces are broken down to their monouculeotides by an exonuclease, which thus serves a scavenging function.

Mature 5S rRNA is resistant to this enzyme. The precursor p5B is also split by RNase M5 into a 22 nucleotide 5' terminal piece, a mature 58 mRNA and some smaller 3' terminal pieces. In eukaryotes the 808 ribosome consists of 40S and 60S subunits. The 408 subunits contain 16-18S rRNA, and the 60S subunits 25-28S rRNA, S.88 RNAand.5S rRNA.

The genes coding for 16-188 rRNA and 25S-288 rRNA arc arranged in clusters of hundreds to thousands of copies. In eukaryotes the transcribed rRNA is 45S rRNA in mammals and 36-38S rRNA in lower eukaryotes, with molecular weights of 4.5 million and 2.6-2.8 million, respectively. There are currently two schemes for the possible arrangement of 28S and 18S segments within the 458 rRNA precursor; and the processing of the precursor rRNA.

According to one scheme 458 rRNA has the general form: 5'P-28s rRNA - spacer - 18SRNA-spacer-3' OH. This transcriptional RNA has a life time of about 15 minutes during which methylation of the ribose moiety in the 288 and 188 regions takes place. An endonuclease cleaves 458 rRNA into 418 and 208components.

The cleavage takes place towards the 5' side of the 208 rRNA components. 41S rNA is degraded through 36S and 328 stages to mature 28S rRNA. During the three steps cleavage takes place in the non-methylated spacer segment by exonucleases. Similarly degration of the non methylated spacer region of the 20S component yields 18S rRNA.

According to the second scheme the transcribed spacer sequence is situated at the 5'P end of 45S rRNA and the 28S com-ponent at the 3'OH end. The bulk of the experimental evidence supports this scheme which would also unify transcription in prokaryotes and eukaryotes. Processing occurs almost entirely within the nucleolus in eukaryote cells. The number of nucleoli within a cell depends on the extent to which the rRNA genes within a chromosome are dispersed.

45S transcriptional rRNA undergoes cleavage at four sites. Cleavage at site I removes the transcribe4 spacer sequence at the 5'P end. In primitive eukaryotes it may occur before transcription is completed. The order of cleavage at sites 2 and 3 varies in different species. Cleavage at site 2 separates 18S rRNA.

A segment of 140 nucleotides, with a sedimentation coefficient of 5.8S (formerly 78), situated between sites 3 and 4, remains covalently linked to the segment formed by.cleavage 3. In yeast tRNAs about 10% of 5.88 RNA is extended at the 5' end by 6-7 nucleotides. Cleavage 4 results in final trimming of the large rRNA segment. 58 rRNA is transcribed separately.

Analyses of the terminal sequences of 16-18S rRNA from

many organisms reveal that the 3' end is almost always A-OH and the 5' end is pU. As many as 8 nucleotides at the 3' end are apparently uniform. They may be involved in the binding of mature rRNA to mRNA. Similarly 6 or more nucleotides may be uniform at the 5' end.

Nucleoside modifications take place on the initial transcript 45S rRNA in regions that give rise to mature 285 and 18S rRNAS. A few base methylations (6 in 18S rRNA) occur at later stages. This late methylation is characteristic of prokaryotes. Mefhy1ation seems to be essential for cleavage.

CLOSURE OF RRNA

Bacterial genome projects usually comprise three well-defined stages. The first one is the construction of high quality random shotgun libraries with varying insert sizes. Following this, the next task is typically to generate a deep coverage of the genome, with random shotgun sequences. This step generates the bulk of the sequences that are used to compile the finished genome sequence, and it is in many ways the heart of the project. Shotgun sequences are then assembled into sequence contigs, each representing a separate, non-overlapping portion of the genome.

The resulting assembly is then converted into a continuous genome sequence through a more laborious and time-consuming finishing phase. Due to the small size and low complexity of bacterial genomes, a genome sequence is only considered finished when all bases meet an acceptable quality level and the individual sequences are assembled into a single continuous consensus sequence. During the finishing phase of a bacterial genome project, the overall sequence quality is improved and sequence data are generated to close gaps between existing contigs.

Finishing can take a variety of forms, but it almost always involves the generation of sequences from large insert clones that have been demonstrated to span virtual gaps in the shotgun assembly, and by the use of alternative strategies to close real gaps when the corresponding DNA fragment has not appeared in any of the libraries generated for the sequencing project. The occurrence of real gaps is often associated with statistical cloning fluctuation of the shotgun model, repetitive regions in the genome (for example ribosomal operons), and with sequences refractory to the cloning system used. Strategies for real gap closure that have been used so far include the use of alternative cloning vectors, combinatory PCR, with primers derived from contig ends, subtractive hybridization, physical mapping, and direct sequencing of genomic DNA.

The time required for finishing depends on the nature of the genome and most particularly on the structure and frequency of repetitive sequences, but this phase easily outlasts that of the generation of the shotgun sequences and is indeed one of the key rate-limiting steps in bacterial genome sequencing.

Here we describe the assembly process and finishing phase of the *Chromobacterium violaceum* genome project. *C. violaceum* is a free-living, gram-negative bacteria that is highly abundant in the water and banks of the Negro River in the Brazilian Amazon, and produces a violacein pigment with antimicrobial activity against some important pathogens as well as antiviral and anticancer activity. The sequencing and analysis of the *C. violaceum* genome were entirely executed by the Brazilian National Genome Sequencing Consortium.

A random shotgun strategy was used, and the resulting sequences were assembled into 57 contigs. These were then organized into 19 scaffolds, using the information from shotgun and cosmid clones; sequences of which were located in different contigs.

Forty-seven virtual gaps within the scaffolds were closed by whole insert sequencing of the corresponding shotgun/cosmid clones. Eighteen real gaps were, for the most part, closed by applying the PCR-assisted contig extension (PACE), followed by specific confirmatory and oriented combinatory PCR. PACE involves the generation of stepwise extensions from the ends of contigs by PCR, until the closure of individual gaps is achieved. This methodology has proven to be especially useful for extending the multicopy ribosomal operons, and has greatly accelerated the finishing phase of the C. *violaceum* genome project.

METHODS

Pace Methodology

PACE was developed as a two-step PCR strategy, with nested primers derived from contig ends, as described by Carraro et al., 2003. Briefly, a set of 96 primers was randomly selected, with no reference to their precise sequence. Pairs of outward-facing specific nested primers were then designed, approximately 140 bp from contig ends and 40 bp apart from each other.

Specific primers were checked for alignment to a single position of the genome with the FASTA programme. Secondary structures and dimer formation were verified using the Oligo Tech software. PCR was then performed in 96-well plates, using 80 ng genomic DNA as templates for the first reaction and 1 ìl of a 1:100 dilution of the first reaction as templates in a subsequent nested reaction.

Specific PCR

All contig extensions were confirmed by specific PCR, followed by direct sequencing of the resulting PCR products. Reaction mixtures for specific PCR contained 80 ng genomic DNA, 250 ìM dNTP, 1.5 mM MgCl2, 1 U Platinum High Fidelity *Taq* DNA polymerase (Invitrogen, Carlsbad, CA, USA), and 10 ìM of each specific primer in a final volume of 20 ìl. Reactions were carried out at 94°C for 30 s, 60°C for 30 s, and 68°C for 2 min from 35 cycles. An initial cycle of 94°C for 2 min, and a final extension at 68°C for 6 min was used.

Combinatory PCR

For ribosomal operon orientation, primers were designed in the flanking regions of the 16S and 5S rRNA genes (outside of the operon repeat unit) and PCRs were undertaken in a combinatory way. The expected fragment sizes were between 5.2 and 5.8 kb, depending on the annealing position of the primers in the flanking region. PCRs were performed in a final volume of 20 ìl, containing 1.5 mM MgC12, 300 ìM dNTPs, 2 U Platinum High Fidelity *Taq* DNA polymerase (Invitrogen) and 15 ìM of each flanking primer. Reactions were carried out at 94°C for 30 s, 60°C for 30 s and 68°C for 5 min, for 35 cycles. Initial denaturation step at 94°C for 2 min, and final extension step at 68°C for 6 min, was used.

Two almost identical versions of the rRNA operon were identified during the assembly phase. The difference between the two copies resides in a 100-bp insertion in the intergenic region between the 16S and ILE genes and a 74-bp insertion between the ILE and ALA genes (copy A - contains 100 bp-16S/ILE, 74 bp- ILE/ALA and copy B - does not contain 100 bp-16S/ILE, 74 bp-ILE/ALA). In order to position correctly the two versions in the genome, a primer located 150 bp downstream of the 100-bp insertion (in the ILE gene) was used in a combinatory way in amplification reactions, together with one of the eight specific primers corresponding to the 16S rRNA flanking region.

PCR reactions were carried out as described above, except for modifications in the cycling parameters (94°C for 30 s, 60°C for 30 s and 68°C for 2 min). Sequence specificity was checked by BLASTN analysis of the two fragment extremities against the two available ribosomal operon copies. A fragment was considered specific to one of the two copies if the high quality sequence portion aligned with the corresponding copy with at least 95% identity.

Product Analysis

Three to five microliters from each PCR were loaded onto a 1% ethidium bromidestained agarose gel. Single PACE products, or fragments of expected size in the case of confirmatory or combinatory PCRs, were purified with the QIAquickTM PCR Purification Kit and sequenced directly using the same specific primer used for amplification on an ABI PrismR 3100 DNA sequencer. High-quality sequences with more than 300 bp, and Phred quality greater than 20, were analyzed for specificity. Sequence specificity was checked by BLASTN analysis against the available genome assembly, and a fragment was considered specific if at least 25 bp of the high quality sequence aligned with at least 95% identity with the end of the corresponding contig.

Chromobacterium Violaceum Genome Assembly

The sequencing and analysis of the *C. violaceum* genome were entirely executed by the Brazilian National Genome Sequencing Consortium, comprising 25 sequencing laboratories, one bioinformatics centre, and three coordination laboratories, distributed throughout Brazil. In the initial phase of the project, random shotgun sequencing produced approximately 80,000 high quality reads, generated from both ends of pUC18 clones, with insert sizes raging from 2.0 to 4.0 kb. Additionally, both ends of 3,350 cosmid clones, with an average insert size of 40 kb, were also sequenced, providing a validation check of the final assembly.

The shotgun sequences, corresponding to approximately 13-fold genome coverage, were assembled into 57 contigs. These shotgun contigs were then organized into 19 scaffolds, using the information from shotgun and cosmid clones, the end sequences of which were located in different contigs. Forty-seven virtual gaps within the 19 scaffolds were closed by whole insert sequencing of the corresponding shotgun/cosmid clones. Points of genome assembly instability and regions of low quality sequences were identified using the Autofinisher Programme, and were resolved by re-sequencing of the selected clones. Real gaps that did not involve the rRNA gene (n = 18) were mainly closed by applying PACE, as detailed in Carraro et al. Here, we relate the methodology used to close rRNA-related gaps and to correctly position the eight ribosomal operon units in the *C. violaceum* genome assembly.

Chromobacterium Violaceum Genome Assembly

After the 16S PACE protocol, the genome assembly was composed of eight unoriented contigs, ending either with 16S rRNA or 5S rRNA sequences. To assemble the contigs in the correct order, we used a combinatory PCR strategy, with primers derived from the non-repetitive genomic region flanking the 16S and 5S rRNA gene. PCR fragments of expected size were sequenced from both extremities, and high quality sequences were aligned to the genome assembly. The orientation of the contigs was confirmed if at least 100 bp of the sequenced fragment aligned with the end of the corresponding contigs, outside the common ribosomal operon sequence. The whole genome assembly revealed a slight difference between the sequences, corresponding to the ribosomal operons. Two almost identical versions were identified.

The difference between the two versions resided in a 100-bp insertion between the 16S and ILE genes, and 74 bp between the ILE and ALA genes. In order to correctly place the two versions (A and B), in the eight already-oriented ribosomal operon copies, a primer located 150 bp downstream of the 100-bp insertion was used in a combinatory way, together with one of the eight specific primers corresponding to the 16S rRNA-flanking region. Using this strategy, we found out that most (6/8) of the rRNA operon copies contained the 100-bp and 74-bp insertion (copy A). These results were independently confirmed by the finding of a larger number of reads in the genome assembly in which the 100-bp insertion was not present.

After the initial assembly of shotgun reads, the *C. violaceum* genome sequence was organized into 19 scaffolds. The genome sequence assembly was mainly made difficult by the eight almost identical rRNA operon copies dispersed throughout the genome. Using PACE, we were able to generate contig extensions with an average of 1 kb in length from all contigs, which closed the majority of gaps in a single round of experimentation. In addition, the PACE methodology proved to be extremely useful for extending the multicopy ribosomal operons. The surprisingly high success rate of our approach can be attributed to deep shotgun coverage and the small size of the gaps in the *C. violaceum* genome assembly. However, deep shotgun coverage is not a pre-requisite for using PACE. Analyses made by our group demonstrated that the methodology can be applied at early stages of the bacterial genome assembly, drastically reducing the time and cost of the finishing phase.

Usually the finishing phase of a genome project takes between 50 to 60% of the total time of the project. In the case of the *C. violaceum* genome project we used approximately 30% of the total time required to conclude the project, corresponding to a reduction of at least 40% of the estimated time. The importance of PACE is that, at a pre-determined point in the shotgun sequencing, primers can be generated and contig extensions obtained with no requirement for a one-by-one analysis of the gaps. Using successive PACE reactions, it is possible to close gaps in bacterial genomes in a stepwise fashion, regardless of their size. Based on the experience accumulated in the *C. violaceum* genome project we strongly recommend the use of the PACE methodology in the finishing phase of bacterial genome projects.

Chapter 10

Protein Synthesis

ONE-GENE-ONE-PROTEIN

The study of infectious disease gave rise to the field of molecular biology. Biochemical reactions are controlled by enzymes, and often are organized into chains of reactions known as metabolic pathways. Loss of activity in a single enzyme can inactivate an entire pathway.

Archibald Garrod, in 1902, first proposed the relationship through his study of alkaptonuria and its association with large quantities "alkapton". He reasoned unaffected individuals metabolized "alkapton" (now called homogentistic acid) to other products so it would not buildup in the urine. Garrod suspected a blockage of the pathway to break this chemical down, and proposed that condition as "an inborn error of metabolism". He also discovered alkaptonuria was inherited as a recessive Mendelian trait.

George Beadle and Edward Tatum during the late 1930s and early 1940s established the connection Garrod suspected between genes and metabolism. They used X rays to cause mutations in strains of the mold *Neurospora.* These mutations affected a single genes and single enzymes in specific metabolic pathways. Beadle and Tatum proposed the "one gene one enzyme hypothesis" for which they won the Nobel Prize in 1958.

Since the chemical reactions occurring in the body are mediated by enzymes, and since enzymes are proteins and thus heritable traits, there must be a relationship between the gene and proteins. George Beadle, during the 1940s, proposed that mutant eye colors in *Drosophila* was caused by a change in one protein in a biosynthetic pathway.

In 1941 Beadle and coworker Edward L. Tatum decided to examine step by step the chemical reactions in a pathway. They used *Neurospora crassa* as an experimental organism. It had a short life-cycle and was easily grown. Since it is haploid for much of its life cycle, mutations would be immediately expressed. The meiotic products could be easily inspected. Chromosome mapping studies on the organism facilitated their work. *Neurospora* can be grown on a minimal medium, and it's nutrition could be studied by its ability to metabolize sugars and other chemicals the scientist could add or delete from the mixture of the medium. It was able to synthesize all of the amino acids and other chemicals needed for it to grow, thus mutants in synthetic pathways would easily show up.

X-rays induced mutations in *Neurospora,* and the mutated spores were placed on growth media enriched with all essential amino acids. Crossing the mutated fungi with non-mutated forms produced spores which were then grown on media supplying only one of the 20 essential amino acids.

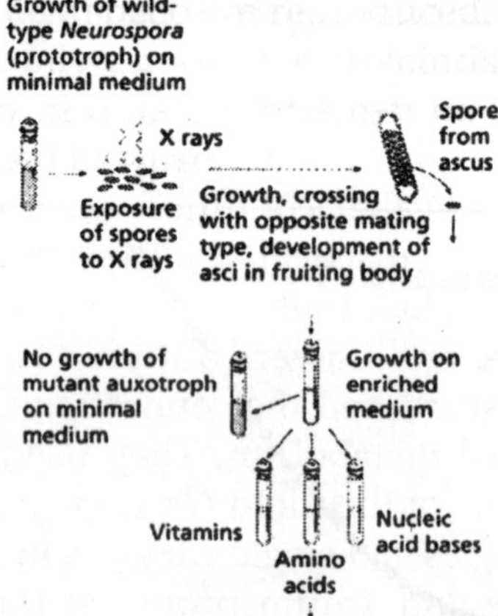

If a spore lacked the ability to synthesize a particular amino acid, such as Pro (proline), it would only grow if the Proline was in the growth medium. Biosynthesis of amino acids (the building blocks of proteins) is a complex process with many chemical reactions mediated by enzymes, which if mutated would shut down the pathway, resulting in no-growth. Beadle and Tatum proposed the "one gene one enzyme" theory.

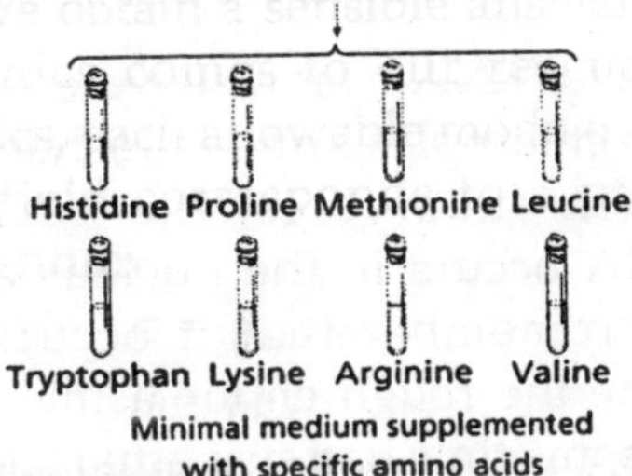

Fig. The Beadle and Tatum Experiment that Suggested the One Gene One Enzyme Hypothesis

One gene codes for the production of one protein. "One gene one enzyme" has since been modified to "one gene one polypeptide" since many proteins (such as hemoglobin) are made of more than one polypeptide.

THE STRUCTURE OF HEMOGLOBIN

Linus Pauling used electrophoresis to separate hemoglobin molecules. Sickle-cell anemia (h) is a recessive allele in which a defective hemoglobin is made, ultimately causing pain and death to those individuals homozygous recessive for the trait.

Pauling reasoned that if Beadle and Tatum were correct, there should be a slight (but detectable) difference between the structure of a normal (HH) and sickle cell (hh) hemoglobin due to genetic differences. Heterozygotes (Hh, also sampled by Pauling) make both normal and "sickle cell" hemoglobins. Later, Vernon Ingram discovered that the normal and sickle-cell hemoglobins differ by only 1 (out of a total of 300) amino acids.

VIRUSES CONTAIN DNA

The coats of viruses act as antigens, initiating an antigen-specific antibody response. Remember that vaccines work by either prompting the immune system to make antibodies or by supplying antibodies. If a virus (or anything else for that matter) mutates its antigens, the immune system is forever playing catch-up.

RNA LINKS THE INFORMATION IN DNA

Ribonucleic acid (RNA) was discovered after DNA. DNA, with exceptions in chloroplasts and mitochondria, is restricted to the nucleus (in eukaryotes, the nucleoid region in prokaryotes). RNA occurs in the nucleus as well as in the cytoplasm (also remember that it occurs as part of the ribosomes that line the rough endoplasmic reticulum).

Scientists for some time had suspected such a link between DNA and proteins. Cells of developing embryos contain high levels of RNA. Rapidly growing *E. coli* has half its mass as ribosomes. Ribosomes are 2/3 RNA (a type of RNA known as ribosomal RNA or rRNA) and 1/3 protein. RNA is synthesized from viral DNA in an infected cell before protein synthesis begins. Some viruses, for example Tobacco Mosaic Virus (TMV) have RNA in place of DNA. If RNA extracted from a virus was injected into a host cell the cell began to make new viruses. Clearly RNA was involved in protein synthesis.

Crick's central dogma. Information flow (with the exception of reverse transcription) is from DNA to RNA via the process of transcription, and thence to protein via translation. Transcription is the making of an RNA molecule off a DNA template. Translation is the construction of an amino acid sequence (polypeptide) from an RNA molecule. Although originally called dogma, this idea has been tested repeatedly with almost no exceptions to the rule being found (save retroviruses).

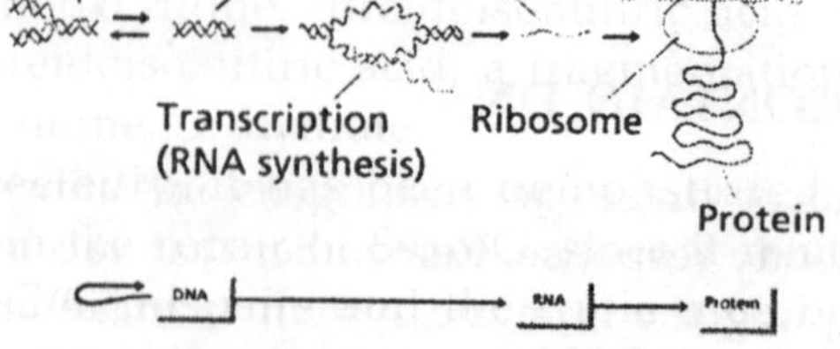

Replication
DNA
mRNA
Translation (protein synthesis)
Transcription (RNA synthesis)
Ribosome
Protein
DNA
RNA
Protein

Fig. The Central Dogma

The code consists of at least three bases, according to astronomer George Gamow. To code for the 20 essential amino acids a genetic code must consist of at least a 3-base set (triplet) of the 4 bases. If one considers the possibilities of arranging four things 3 at a time (4X4X4), we get 64 possible code words, or codons (a 3-base sequence on the mRNA that codes for either a specific amino acid or a control word).

The genetic code was broken by Marshall Nirenberg and Heinrich Matthaei, a decade after Watson and Crick's work. Nirenberg discovered that RNA, regardless of its source organism, could initiate protein synthesis when combined with contents of broken E. coli cells. By adding poly-U to each of 20 test-tubes (each tube having a different "tagged" amino acid) Nirenberg and Matthaei were able to determine that the codon UUU (the only one in poly-U) coded for the amino acid phenylalanine.

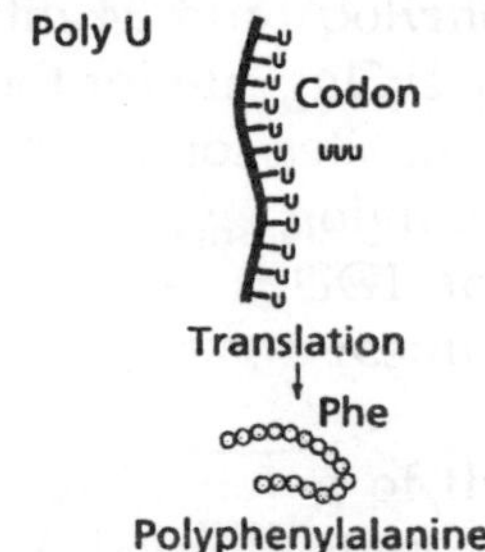

Fig. Steps in Breaking the Genetic Code: the Deciphering of a Poly-U mRNA

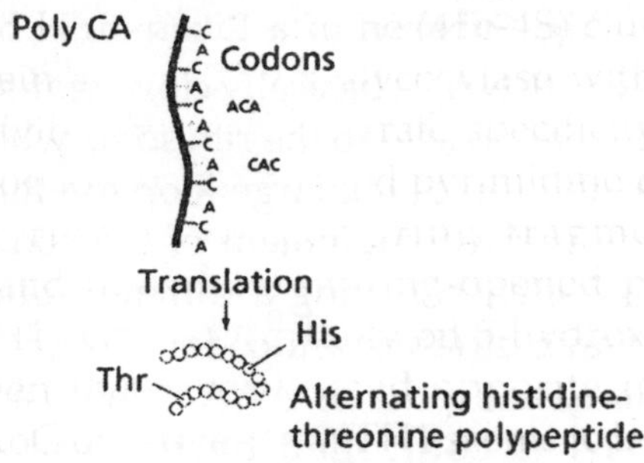

Fig. Deciphering the Code: Poly CA

Likewise, an artificial mRNA consisting of alternating A

and C bases would code for alternating amino acids histidine and threonine. Gradually, a complete listing of the genetic code codons was developed.

PROTEIN SYNTHESIS

Prokaryotic gene regulation differs from eukaryotic regulation, but since prokaryotes are much easier to work with, we focus on prokaryotes at this point. Promoters are sequences of DNA that are the start signals for the transcription of mRNA. Terminators are the stop signals. mRNA molecules are long (500- 10,000 nucleotides).

Ribosomes are the organelle (in all cells) where proteins are synthesized. They consist of two-thirds rRNA and one-third protein. Ribosomes consist of a small and larger (50S) subunits. The length of rRNA differs in each. The 30S unit has 16S rRNA and 21 different proteins. The 50S subunit consists of 5S and 23S rRNA and 34 different proteins. The smaller subunit has a binding site for the mRNA. The larger subunit has two binding sites for tRNA.

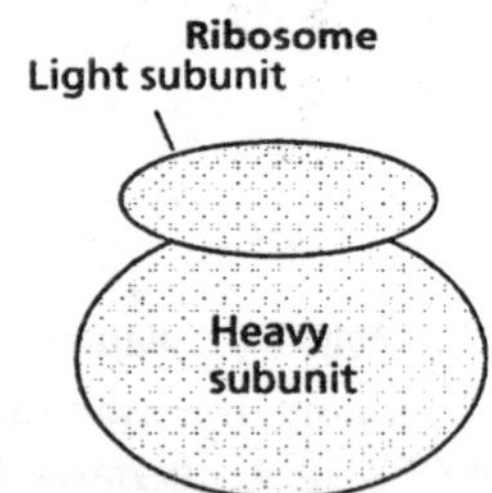

Fig. Subunits of a Ribosome

Transfer RNA (tRNA) is basically cloverleaf-shaped. tRNA carries the proper amino acid to the ribosome when the codons call for them. At the top of the large loop are three bases, the anticodon, which is the complement of the codon. There are 61 different tRNAs, each having a different binding site for the amino acid and a different anticodon. For the codon UUU, the complementary anticodon is AAA. Amino acid linkage to the proper tRNA is controlled by the aminoacyl-tRNA synthetases. Energy for binding the amino acid to tRNA comes from ATP conversion to adenosine monophosphate (AMP).

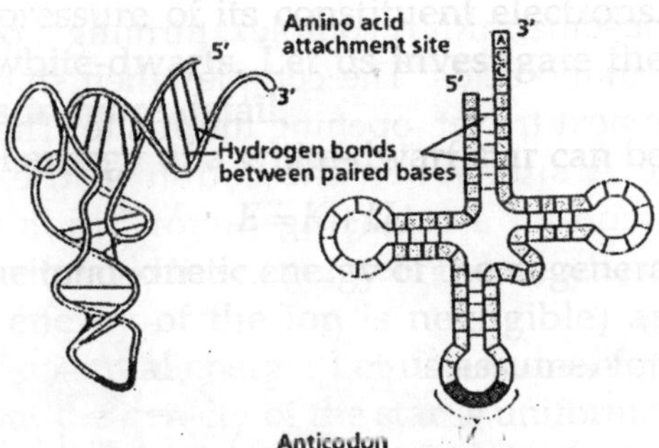

Fig. Two Models of tRNA

Translation is the process of converting the mRNA codon sequences into an amino acid sequence. The initiator codon (AUG) codes for the amino acid N-formylmethionine (f-Met). No transcription occurs without the AUG codon. f-Met is always the first amino acid in a polypeptide chain, although frequently it is removed after translation. The intitator tRNA/mRNA/small ribosomal unit is called the initiation complex. The larger subunit attaches to the initiation complex. After the initiation phase the message gets longer during the elongation phase.

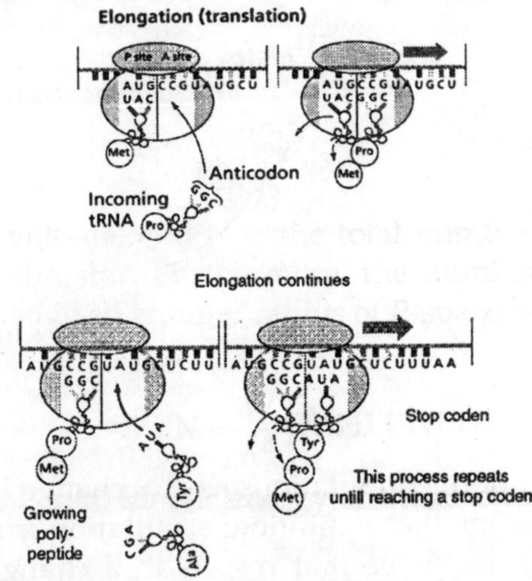

Fig. Translation

New tRNAs bring their amino acids to the open binding

site on the ribosome/mRNA complex, forming a peptide bond between the amino acids. The complex then shifts along the mRNA to the next triplet, opening the A site. The new tRNA enters at the A site. When the codon in the A site is a termination codon, a releasing factor binds to the site, stopping translation and releasing the ribosomal complex and mRNA.

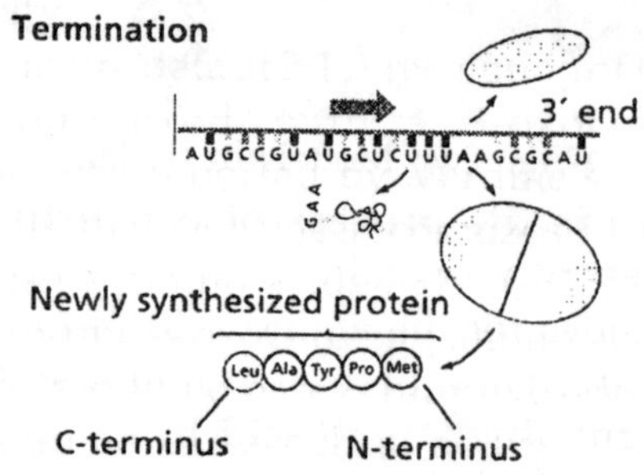

Fig. Termination

Often many ribosomes will read the same message, a structure known as a polysome forms. In this way a cell may rapidly make many proteins.

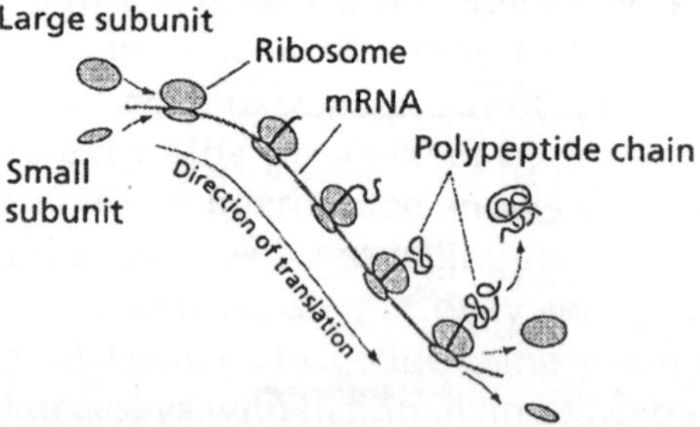

Fig. Many Ribosomes Translating the same Message, a Polysome

MUTATIONS REDEFINED

We earlier defined mutations as any change in the DNA. We now can refine that definition: a mutation is a change in the DNA base sequence that results in a change of amino acid(s) in the polypeptide coded for by that gene. Alleles are alternate sequences of DNA bases (genes), and thus at the molecular level the products of alleles differ (often by only a single amino acid, which can have a ripple effect on an organism by changing).

Addition, deletion, or addition of nucleotides can alter the polypeptide. Point mutations are the result of the substitution of a single base. Frame-shift mutations occur when the reading frame of the gene is shifted by addition or deletion of one or more bases. With the exception of mitochondria, all organisms use the same genetic code. Powerful evidence for the common ancestry of all living things.

PROTEIN BIOSYNTHESIS

Protein biosynthesis (synthesis) is the process in which cells build proteins. The term is sometimes used to refer only to protein translation but more often it refers to a multi-step process, beginning with amino acid synthesis and transcription which are then used for translation. Protein biosynthesis, although very similar, differs between prokaryotes and eukaryotes.

AMINO ACID SYNTHESIS

Amino acids are the monomers which are polymerized to produce proteins. Amino acid synthesis is the set of biochemical processes (metabolic pathways) which build the amino acids from carbon sources like glucose. Not all amino acids may be synthesised by every organism, for example adult humans have to obtain 8 of the 20 amino acids from their diet.

TRANSCRIPTION

Transcription is the process by which an mRNA template, encoding the sequence of the protein in the form of a trinucleotide code, is transcribed from the genome to provide a template for translation. Transcription copies the template from one strand of the DNA double helix, called the template strand. Transcription can be divided into 3 stages: Initiation, Elongation and Termination, each regulated by a large number of proteins such as transcription factors and coactivators that ensure the correct gene is transcribed in response to appropriate signals. The DNA strand is read in the 3' to 5' direction and the mRNA is transcribed in the 5' to 3' direction by the RNA polymerase.

TRANSLATION

The synthesis of proteins is known as translation. Translation occurs in the cytoplasm where the ribosomes are located. Ribosomes are made of a small and large subunit which surrounds the mRNA.

In translation, messenger RNA (mRNA) is decoded to produce a specific polypeptide according to the rules specified by the genetic code. This uses an mRNA sequence as a template to guide the synthesis of a chain of amino acids that form a protein. Translation is necessarily preceded by transcription. Translation proceeds in four phases: activation, initiation, elongation and termination (all describing the growth of the amino acid chain, or polypeptide that is the product of translation).

In activation, the correct amino acid (AA) is joined to the correct transfer RNA (tRNA). While this is not technically a step in translation, it is required for translation to proceed. The AA is joined by its carboxyl group to the 3' OH of the tRNA by an ester bond. When the tRNA has an amino acid linked to it, it is termed "charged". Initiation involves the small subunit of the ribosome binding to 5' end of mRNA with the help of initiation factors (IF), other proteins that assist the process.

Elongation occurs when the next aminoacyl-tRNA (charged tRNA) in line binds to the ribosome along with GTP and an elongation factor. Termination of the polypeptide happens when the A site of the ribosome faces a stop codon (UAA, UAG, or UGA).

When this happens, no tRNA can recognize it, but releasing factor can recognize nonsense codons and causes the release of the polypeptide chain. The capacity of disabling or inhibiting translation in protein biosynthesis is used by antibiotics such as: anisomycin, cycloheximide, chloramphenicol, tetracycline, streptomycin, erythromycin, puromycin etc.

EVENTS FOLLOWING PROTEIN TRANSLATION

The events following biosynthesis include post-translational modification and protein folding. During and after synthesis, polypeptide chains often fold to assume, so called, native secondary and tertiary structures. This is known as *protein folding*. Many proteins undergo *post-translational modification*. This may include the formation of disulfide bridges or attachment of any of a number of biochemical functional groups, such as acetate, phosphate, various lipids and carbohydrates. Enzymes may also remove one or more amino acids from the leading (amino) end of the polypeptide chain, leaving a protein consisting of two polypeptide chains connected by disulfide bonds.

PEPTIDE SYNTHESIS

In organic chemistry, peptide synthesis is the production of peptides, which are organic compounds in which multiple amino acids bind via peptide bonds which are also known as amide bonds. The biological process of producing long peptides (proteins) is known as protein biosynthesis.

Chemistry

Peptides are synthesized by coupling the carboxyl group or C-terminus of one amino acid to the amino group or N-terminus of another.

Liquid-phase Synthesis

Liquid-phase peptide synthesis is a classical approach to peptide synthesis. It has been replaced in most labs by solid-phase synthesis. However, it retains usefulness in large-scale production of peptides for industrial purposes.

Solid-phase Synthesis

Solid-phase peptide synthesis (SPPS), pioneered by Robert Bruce Merrifield, resulted in a paradigm shift within the peptide synthesis community. It is now the accepted method for creating peptides and proteins in the lab in a synthetic manner. SPPS allows the synthesis of natural peptides which are difficult to express in bacteria, the incorporation of unnatural amino acids, peptide/protein backbone modification, and the synthesis of D-proteins, which consist of D-amino acids.

Small solid beads, insoluble yet porous, are treated with functional units ('linkers') on which peptide chains can be built. The peptide will remain covalently attached to the bead until cleaved from it by a reagent such as trifluoroacetic acid. The peptide is thus 'immobilized' on the solid-phase and can be retained during a filtration process, whereas liquid-phase reagents and by-products of synthesis are flushed away.

The general principle of SPPS is one of repeated cycles of coupling-deprotection. The free N-terminal amine of a solid-phase attached peptide is coupled to a single N-protected amino acid unit. This unit is then deprotected, revealing a new N-terminal amine to which a further amino acid may be attached.

The overwhelmingly important consideration is to generate extremely high yield in each step. For example, if each coupling step were to have 99% yield, a 26-amino acid peptide would be synthesized in 77% final yield (assuming 100% yield in each deprotection); if each step were 95%, it would be synthesized in 25% yield. Thus each amino acid is added in major excess (2~10x) and coupling amino acids together is highly optimized by a series of well-characterized agents.

There are two majorly used forms of SPPS — Fmoc and Boc. Unlike ribosome protein synthesis, solid-phase peptide synthesis proceeds in a C-terminal to N-terminal fashion. The N-termini of amino acid monomers is protected by these two groups and added onto a deprotected amino acid chain.

Automated synthesizers are available for both techniques, though many research groups continue to perform SPPS manually. SPPS is limited by yields, and typically peptides and proteins in the range of 70~100 amino acids are pushing the limits of synthetic accessibility. Synthetic difficulty also is sequence dependent; typically amyloid peptides and proteins are difficult to make. Longer lengths can be accessed by using native chemical ligation to couple two peptides together with quantitative yields.

t-Boc Solid-phase Peptide Synthesis

When Merrifield invented SPPS in 1963, it was according to the t-Boc method. t-Boc (or Boc) stands for (t)ert-(B)ut(o)xy(c)arbonyl. To remove Boc from a growing peptide chain, acidic conditions are used (usually neat TFA). Removal of side-chain protecting groups and the peptide from the resin at the end of the synthesis is achieved by incubating in hydrofluoric acid (which can be dangerous or even deadly); for this reason Boc chemistry is generally disfavored.

Also, HF cleavage needs to be done in special fume hoods using specialized equipment. However for complex syntheses Boc is favourable. When synthesizing nonnatural peptide analogs which are base-sensitive (such as depsipeptides), Boc is necessary.

Fmoc Solid-phase Peptide Synthesis

This method was introduced by Carpino in 1972 and further applied by Atherton in 1978. Fmoc stands for 9*H*-(f)luoren-9-yl(m)eth(o)xy(c)arbonyl which describes the Fmoc protecting group, first described as a protecting group by Carpino in 1970. To remove an Fmoc from a growing peptide chain, basic conditions (usually 20% piperidine in DMF) are used.

Removal of side-chain protecting groups and peptide from the resin is achieved by incubating in trifluoroacetic acid (TFA), deionized water, and triisopropylsilane. Fmoc deprotection is usually slow because the anionic nitrogen produced at the end is not a particularly favorable product, although the whole process is thermodynamically driven by the evolution of carbon dioxide.

The main advantage of Fmoc chemistry is that no hydrofluoric acid is needed. It is therefore used for most routine synthesis.

Bop Spps

The use of BOP reagent was first described by Castro et al in 1975.

SOLID SUPPORTS

The physical properties of the solid support, and the applications to which it can be utilized, vary with the material from which the support is constructed, the amount of crosslinking, as well as the linker and handle being used.

Polystyrene Resin

Polystyrene resin is a versatile resin and it is quite useful in multi-well, automated peptide synthesis, due to its minimal swelling in dichloromethane.

Polyamide Resin

Polyamide resin is also a useful and versatile resin. It seems to swell much more than polystyrene, in which case it may not be suitable for some automated synthesizers, if the wells are too small.

PEG Based Resin

ChemMatrix(R) is a new type of resin which is based on PEG that is crosslinked. ChemMatrix(R) has claimed a high chemical and thermal stability (is compatible with Microwave synthesis) and has shown higher degrees of swellings in acetonitrile, dichloromethane, DMF, N-methyl pyrollidone, TFA and water compared to the polystyrene based resins. ChemMatrix has shown significant improvements to the synthesis of hydrophobic sequences. ChemMatrix is recommended for the synthesis of difficult and long peptides.

PROTECTING GROUPS

Due to amino acid excesses used to ensure complete coupling during each synthesis step, polymerization of amino acids is common in reactions where each amino acid is not protected. In order to prevent this polymerization, protecting groups are used. This adds additional deprotection phases to the synthesis reaction, creating a repeating design flow as follows:

- Protective group is removed from trailing amino acids in a deprotection reaction
- Deprotection reagents washed away to provide clean coupling environment
- Protected amino acids dissolved in a solvent such as dimethylformamide (DMF) are combined with coupling reagents are pumped through the synthesis column
- Coupling reagents washed away to provide clean deprotection environment

Currently, two protective groups (t-Boc, Fmoc) are commonly used in solid-phase peptide synthesis. Their lability is caused by the carbamate group which readily releases CO_2 for an irreversible decoupling step.

t-Boc Protective Group

The t-Boc group was commonly used for protecting the terminal amine of the peptide, requiring the use of more acid stable groups for side chain protection in orthogonal strategies. It retains usefulness in reducing aggregation of peptides during synthesis. Boc groups can be added to amino acids with t-Boc anhydride and a suitable base.

Fmoc Pprotective Group

Fmoc (*9H*-fluoren-9-ylmethoxycarbonyl) is currently a widely used protective group that is generally removed from the N terminus of a peptide in the iterative synthesis of a peptide from amino acid units. The advantage of Fmoc is that it is cleaved under very mild basic conditions (e.g. piperidine), but stable under acidic conditions. This allows mild acid labile protecting groups that are stable under basic conditions, such as Boc and benzyl groups, to be used on the side-chains of amino acid residues of the target peptide. This orthogonal protecting group strategy is common in the art of organic synthesis.

FMOC is preferred over BOC due to ease of cleavage; however it is less atom-economical, as the fluorenyl group is much larger than the tert-butyl group. Accordingly, prices for FMOC amino acids were high until the large-scale piloting of one of the first synthesized peptide drugs, enfuvirtide, began in the 1990s, when market demand adjusted the relative prices of the two sets of amino acids.

Benzyloxy-carbonyl (Z) Group

The first use of (Z) group as protecting groups was done by Max Bergmann who synthesised oligopeptides. Another carbamate based group is the benzyloxy-carbonyl (Z) group. It is removed in harsher conditions: HBr/acetic acid or catalytic hydrogenation. Today it is almost exclusively used for side chain protection.

Alloc Protecting Group

The allyloxycarbonyl (alloc) protecting group is often used to protect a carboxylic acid, hydroxyl, or amino group when an orthogonal deprotection scheme is required. It is sometimes used when conducting on-resin cyclic peptide formation, where the peptide is linked to the resin by a side-chain functional group.

The alloc group can be removed using tetrakis (triphenylphosphine)palladium(0) along with a 37:2:1 mixture of chloroform, acetic acid, and N-methylmorpholine (NMM) for 2 hours. The resin must then be carefully washed 0.5% DIPEA in DMF, 3x10 ml of 0.5% sodium diethylthiocarbamate in DMF, and then 5x10 ml of 1:1 DCM:DMF.

Lithographic Protecting Groups

For special applications like protein microarrays lithographic protecting groups are used. Those groups can be removed through exposure to light.

ACTIVATING GROUPS

For coupling the peptides the carboxyl group is usually activated. This is important for speeding up the reaction. There are two main types of activating groups: carbodiimides and triazolols.

Carbodiimides

These activating agents were first developed. Most common are dicyclohexylcarbodiimide (DCC) and diisopropylcarbodiimide (DIC). Reaction with a carboxylic acid yields a highly reactive O-acyl-urea. During artificial protein synthesis (such as Fmoc solid-state synthesizers), the C-terminus is often used as the attachment site on which the amino acid monomers are added. To enhance the electrophilicity of carboxylate group, the negatively charged oxygen must first be "activated" into a better leaving group. DCC is used for this purpose.

The negatively charged oxygen will act as a nucleophile, attacking the central carbon in DCC. DCC is temporarily attached to the former carboxylate group (which is now an ester group), making nucleophilic attack by an amino group (on the attaching amino acid) to the former C-terminus (carbonyl group) more efficient. The problem with carbodiimides is that they are too reactive and that they can therefore cause racemization of the amino acid.

Triazolols

To solve the problem of racemization, triazolols were introduced. The most important ones are 1-hydroxy-benzotriazole (HOBt) and 1-hydroxy-7-aza-benzotriazole (HOAt). Others have been developed. These substances can react with the O-acylurea to form an active ester which is less reactive and less in danger of racemization. HOAt is especially favourable because of a neighbouring group effect.

Newer developments omit the carbodiimides totally. The active ester is introduced as a uronium or phosphonium salt of a non-nucleophilic anion (tetrafluoroborate or hexafluorophosphate): HBTU, HATU, PyBOP.

SYNTHESIZING LONG PEPTIDES

Stepwise elongation, in which the amino acids are connected step-by-step in turn, is ideal for small peptides containing between 2 and 100 amino acid residues. Another method is fragment condensation, in which peptide fragments are coupled.

Although the former can elongate the peptide chain without racemization, the yield drops if only it is used in the creation of long or highly polar peptides. Fragment condensation is better than stepwise elongation for synthesizing sophisticated long peptides, but its use must be restricted in order to protect against racemization. Fragment condensation is also undesirable since the coupled fragment must be in gross excess, which may be a limitation depending on the length of the fragment.

A new development for producing longer peptide chains is chemical ligation: Unprotected peptide chains react chemoselectively in aqueous solution.

A first kinetically controlled product rearranges to form the amide bond. The most common form of native chemical ligation uses a peptide thioester that reacts with a terminal cystein residue.

MICROWAVE ASSISTED PEPTIDE SYNTHESIS

Although microwave irradiation has been around since the late 1940s, it was not until 1986 that microwave energy was used in organic chemistry. During the end of the 1980s and 1990s, microwave energy was an obvious source for completing chemical reactions in minutes that would otherwise take several hours to days. Through several technical improvements at the end of the 1990s and beginning of the 2000s, microwave synthesizers have been designed to provide both low and high energy pockets of microwave energy so that the temperature of the reaction mixture could controlled.

The microwave energy used in peptide synthesis is of a single frequency providing maximum penetration depth of the sample which is in contrast to conventional kitchen microwaves. In peptide synthesis, microwave irradiation has been used to complete long peptide sequences with high degrees of yield and low degrees of racemization. Microwave irradiation during the coupling of amino acids to a growing polypeptide chain is not only catalyzed through the increase in temperature, but also due to the alternating electromagnetic radiation to which the polar backbone of the polypeptide continuously aligns to.

Due to the this phenomenon, the microwave energy can prevent aggregation and thus increases yields of the final peptide product. Despite the main advantages of microwave irradiation of peptide synthesis, the main disadvantage is the racemization which may occur with the coupling of cysteine and histidine.

A typical coupling reaction with these amino acids are performed at lower temperatures than the other 18 natural amino acids. Another disadvantage is that allyl containing amino acid derivatives cannot be coupled to amino acids using microwave irradiation due to uncontrolled polymerization.

CYCLIC PEPTIDES

AN EXAMPLE OF SOLID PHASE PEPTIDE SYNTHESIS

The following is an outline of the synthetic steps for peptide synthesis on polyamide or polystyrene resin, using the base labile 9*H*-fluoren-9-ylmethoxycarbonyl (Fmoc) protecting group. Using the techniques outlined below, one will obtain a peptide which is capped on the N-terminus with and acetyl group, and on the C-terminus with a primary amide ($CONH_2$).

Setting Up Glassware for Manual Peptide Synthesis

Manual peptide synthesis can be accomplished in a fritted-filter reaction vessel with a three-way valve fitted onto a 1 L side arm vacuum flask by way of a 1-hole stopper. One valve is used to bubble nitrogen, which is first passed through a small column of Drierite, and then into the reaction mixture to agitate the solution and mix reagents. The other valve is used to evacuate excess reaction solutions and wash solvent using a vacuum flask. All glass pieces to be used in Solid-phase synthesis should be treated with a silanizing agent (such as 1-5% dimethyldichlorosilane in DCM) prior to use, to avoid accumulation of static charge, which makes the resin very difficult to handle.

Handling the Resin before, and during Synthesis

The resin is first swelled for 15 minutes in 10 ml of 1:1 DCM:DMF and drained. The resin is also washed with 5x10 ml of 1:1 DCM and DMF after each completed amino acid coupling.

Fmoc-Deprotection

Fmoc deprotection after each amino acid coupling is accomplished using 2 × 10 ml of 20% piperidine in DMF, with N2 agitation for 10 minutes each treatment. The resin is then washed with 5 × 10 ml DMF, followed by 5 × 10 ml of 1:1 DCM:DMF.

An alternate treatment is 1% DBU in DMF; this gentler treatment allows removal of Fmoc groups in the presence of other base-labile moieties.

Monitoring the Progress of Amino Acid Couplings

The progress of amino acid couplings can be followed using ninhydrin, or p-chloranil.

The ninhydrin solution turns dark blue (positive result) in the presence of a free primary amine but is otherwise colorless (negative result).

The p-chloranil solution will turn the resin beads dark black or blue in the presence of a primary amine if acetaldehyde is used as the solvent or in the presence of a secondary amine, if acetone is used instead; the beads remain colorless or pale yellow otherwise.

Testing by Ninhydrin (1)

Add 2 drops of 40% phenol in ethanol, 2 drops of 0.014

mol/L KCN in pyridine, and 4 drops of 5% ninhydrin in ethanol to a microcentrifuge tube along with a spatula tip size sample of resin, then vortex the mixture and heat for 5 minutes at 100°C.

Testing by Chloranil (2)

Add 5 drops of acetone or acetaldehyde, 5 drops of a saturated solution of p-chloranil in toluene, plus a small spatula-tip-size sample of resin to a microcentrifuge tube, then vortex the mixture and allow to stand at room temperature for 5 minutes. Acetone is used for the detection of secondary amines, where acetaldehyde is used for primary amines.

Continuing Peptide Extension: Once the coupling of the amino acid is complete, the resin is washed, the Fmoc group deprotected with piperidine, and the resin washed again to prepare it for the next coupling. This process is repeated until all necessary amino acids have been added.

Acetylating the N-terminus: After the peptide sequence is completed, the N-terminal amine can be acetylated with 2 ml of 1:1 acetic anhydride and triethylamine in 10 ml of 1:1 DCM:DMF for 1 hour or until negative to ninhydrin, and the resin then washed with 5x10 ml of 1:1 DCM:DMF, before the peptide is cleaved from the resin. The N-terminus can also be left as the free amine if required.

Cleaving the Peptide from the Resin

The resin is treated with a cocktail of trifluoroacetic acid (TFA) and cleavage scavenger reagents, such as triisopropylsilane (TIPS), 1,2-ethanedithiol, and water (H_2O). The choice of scavengers is dependent on the amino acid sequence. The resin is then filtered away, and the combined filtrates allowed to stand for 1 hour to ensure removal of the acid labile protecting groups.

Workup of Peptides After Cleavage from the Resin

The TFA is evaporated to dryness (or a heavy oil or glass if it does not solidify) on the rotary evaporator, followed by the addition of 5 ml of diethyl ether to the flask to precipitate the peptide, and remove the bulk of the by-products.

Once the peptide is precipitated with the ether, filter through a sintered glass funnel and redissolve peptide into 60% AcN/ 0.1% TFA. The peptide solution is then frozen in a dry ice/ethanol bath and lyophilised (dried).

The result is a crude peptide which is stable for a number of years. The peptide can then be purified by Ion-exhange and Reverse Phase chromatography. Typically preparative HPLC is used to purify the final product. Mass spectrometry data is obtained to ensure the target peptide was obtained.

Index

D

E

Q

R

S

T

U

V

W

Y